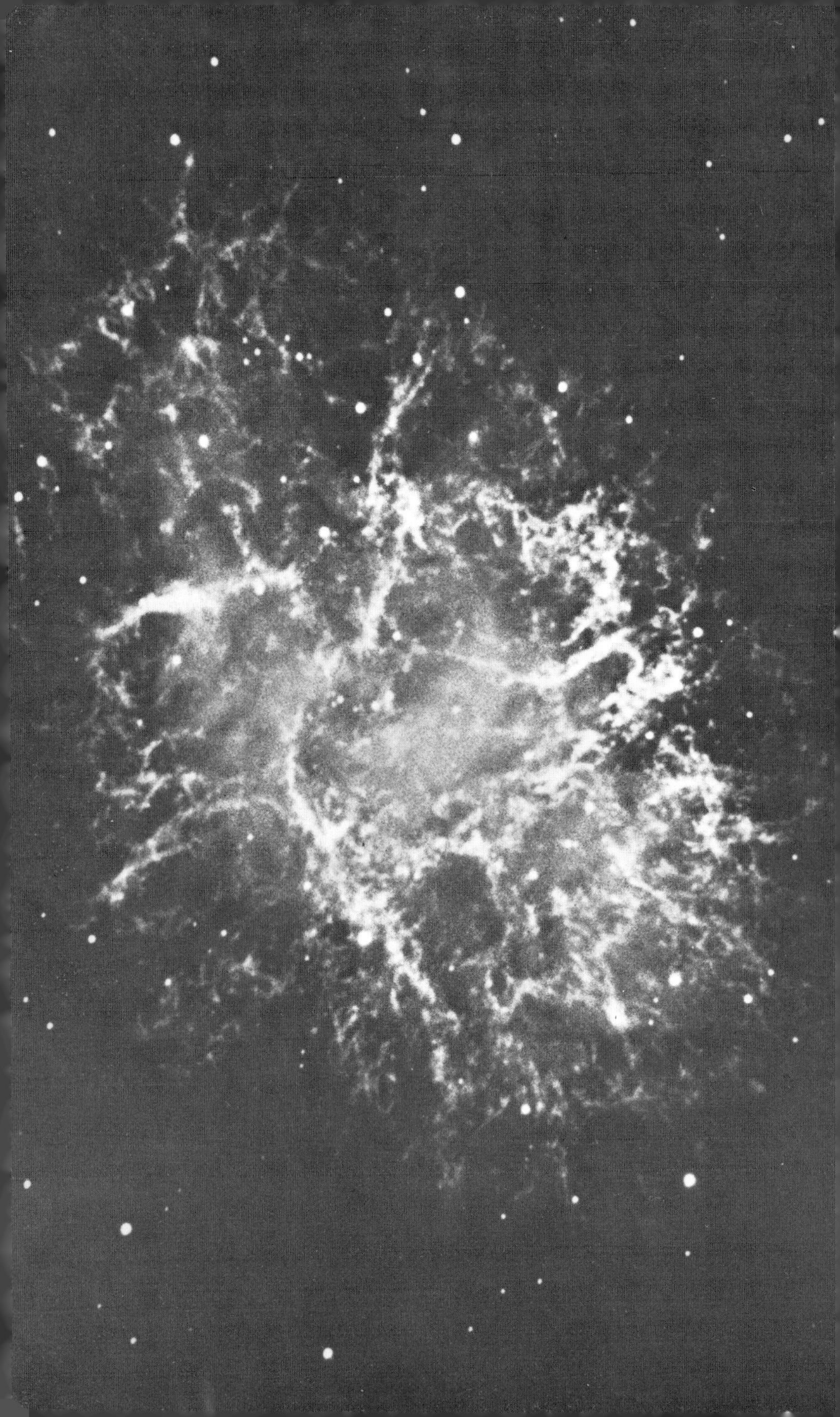

THE UNIVERSE IS A GREEN DRAGON

Cover Photograph: THE ORION NEBULA

"The Orion Nebula glows with the elements ionized by the newly created stars in the center of the cloud. One of these infant stars began shining only two thousand years ago."

Brian Swimme

THE UNIVERSE IS A GREEN DRAGON

A Cosmic Creation Story

Brian Swimme

Bear & Company
Santa Fe, New Mexico

PHOTO INSERT

Photo 1: THE TRIFID NEBULA IN SAGITTARIUS;
Lick Observatory Photograph.

Photo 2: CRAB NEBULA; Lick Observatory Photograph.

Photo 3: THE LAGOON NEBULA; Lick Observatory Photograph.

Photo 4: IC 1318; Lick Observatory Photograph.

Photo 5: THE ORION NEBULA; Lick Observatory Photograph.

Photo 6: ANDROMEDA GALAXY; Lick Observatory Photograph.

Photo 7: THE CALIFORNIA NEBULA IN PERSEUS;
Lick Observatory Photograph.

Photo 8: THE NEBULOUS CLUSTER MESSIER 16;
Lick Observatory Photograph.

ISBN 0-939680-14-9

Library of Congress Card Number 84-072255

Bear & Company, Inc.
P.O. Drawer 2860
Santa Fe, NM 87504

Cover Design: William Field, Santa Fe

Printed in the United States by BookCrafters, Inc.

Video tapes of Briam Swimme presenting themes from *The Universe Is A Green Dragon* are available from Friends of Creation Spirituality, Inc., PO Box 19216, Oakland, CA 94619.

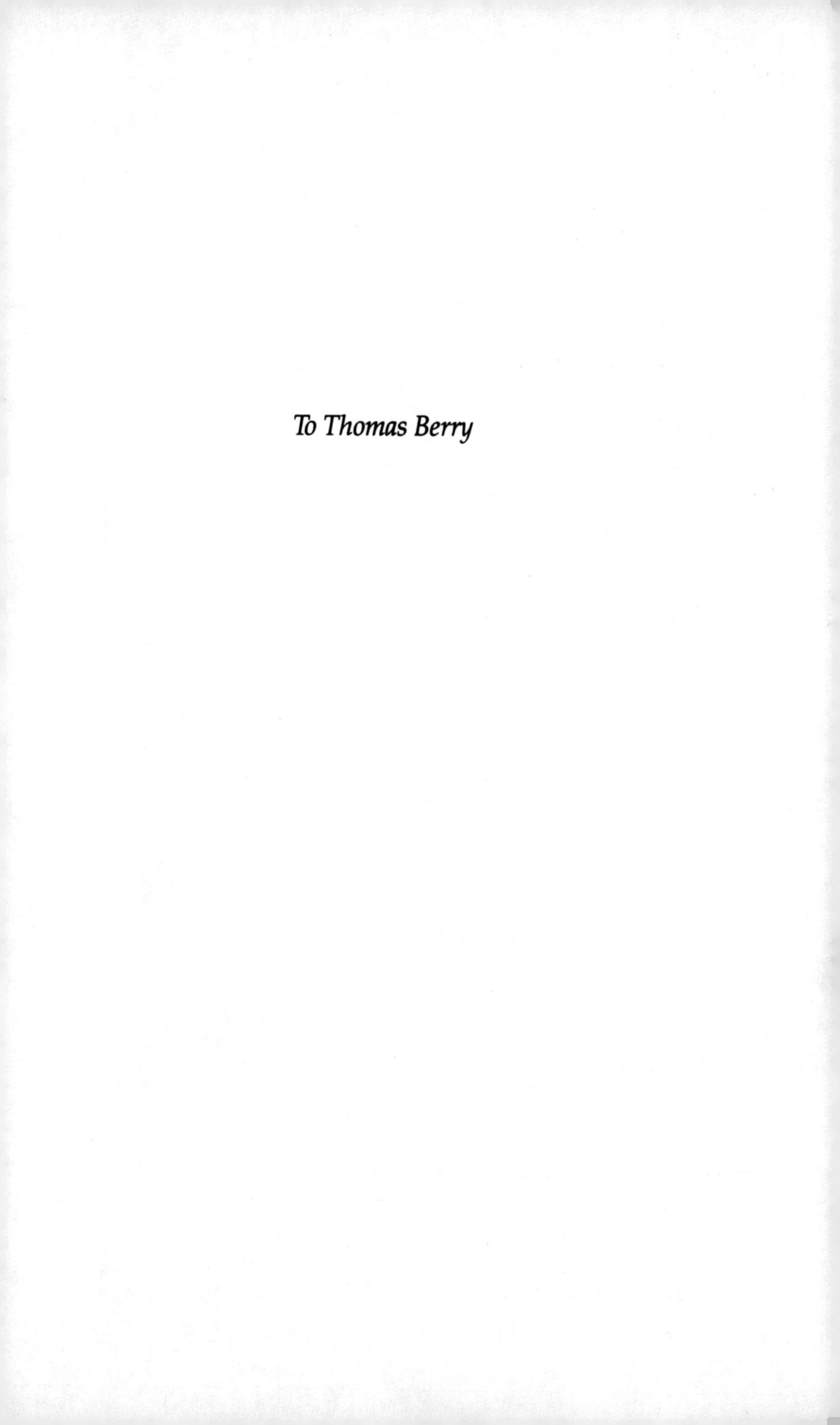

To Thomas Berry

CONTENTS

PROLOGUE

I was presenting some ideas on the new cosmology at a conference in Chicago when suddenly a woman charged out of the audience, upset, eyes flashing as if Athena herself had decided to confront me: "I want you to explain to me why my son isn't taught this in high school. You say scientists have thrown out the materialistic world view. Then why should my son suffer it at all?"

A good question. And it concerns more than our high schools. I used to wonder something similar when I taught mathematics and physics at the university level. I was supposed to introduce students to the universe, the *universe*, but I was not to speak of meaning. Doesn't that seem like a strange assignment?

If you persist in such questioning, the answer is not hard to find. Our modern western civilization began with a kind of cultural schizophrenia. Our scientific enterprise effectively decoupled itself from our humanistic-spiritual traditions at the beginning of the modern period. All for good reasons, yes, but now the neurosis spreads over several continents. Enmeshed in the most terrifying pathology in the history of humanity, we can perhaps dare to ask if this was such a good idea, this splitting up of the universe.

Alert humans could see the danger of our situation from the beginning. Though they could not have predicted the planetary poisons that beset us, nor the threat of annihilation that everyone carries to bed every night of the week, they could see that we were headed for an unhealthy future. Diseased mindscapes only produce diseased landscapes. But there was nothing anyone could do about it. The sciences were effective in their mechanistic formulations and thus became entrenched in mechanism. Our religious tradition carefully retreated into a redemption orientation and concluded Creation was not its concern. Western culture put itself on tracks that led to an inescapable and ever deepening sickness.

But something tremendous is occurring in our time, something with the power to break up this impasse. I mean the radical transformation of our fundamental world view as the cosmic story of our origins and development takes hold in human awareness. When I say *our origins and development*, I mean more than the human species. I mean the origin and development of the universe as a whole. We have discovered something with overwhelming possibilities. The universe can no longer be regarded as a result of chance collisions of materials, nor as a deterministic mechanism. The universe considered as a whole is more like a developing being. The universe has a beginning and is in the midst of its development: a vast cosmic epigenesis. Everything that exists is involved in this emergence—galaxies and stars and planets and light and all living creatures.

How does the deeper understanding empower us? By enabling us to reinvent the human within the new cosmic story. Nothing less will suffice. A new sociological viewpoint or a new psychological theory is inadequate to deal with the magnitude of our concerns. We need to understand the human within the intrinsic dynamics of the Earth. Alienated from the cosmos, imprisoned in our narrow frames of reference, we do not know what we are about as a species. We will discover our larger role only by reinventing the human as a dimension of the emergent universe.

What I present in the pages of this book is the overall picture of the cosmic creation story, told in a single evening's conversation.

I call the two speakers THOMAS and YOUTH. By

THOMAS I want to honor Thomas Berry and the cosmological tradition he celebrates, stretching back from Erich Jantsch and Teilhard de Chardin through Thomas Aquinas to Plato. The idea to present the new creation story in the form of a conversation originated at the Broadway Diner in New York City. I was working my way through a Greek salad, when Thomas Berry suddenly said: "You scientists have this stupendous story of the universe. It breaks outside all previous cosmologies. But so long as you persist in understanding it solely from a quantitative mode you fail to appreciate its significance. You fail to hear its music. That's what the spiritual traditions can provide. Tell the story, but tell it with a feel for its music."

I call the other human YOUTH to remind us that the human species is the youngest, freshest, most immature, newest species of all the advanced life forms in the planet. We have only just arrived. If we can remain resilient, if we can continue our questioning, our developing, our hoping, if we can live in awe and in the depths of wonder, we will continue moving into the only process that now matters—our authentic maturation as a species. It is in this way and only this way that we will enable the Earth to bloom once again.

I: COSMOS AS PRIMARY REVELATION

CREATIVITY: PRIMORDIAL AND PERVASIVE

YOUTH: Why do you say the universe is a green dragon?

THOMAS: I'm a storyteller. Besides, it seems an appropriate way to begin the new story of the cosmos.

YOUTH: But why say it's a green dragon when it obviously isn't?

THOMAS: For several reasons. I call the universe a green dragon to remind us that we will never be able to capture the universe with language.

YOUTH: How can you be certain of that?

THOMAS: Because the universe is a singularity! To speak, you need to compare things. Thus we say that the house is white, not brown. Or that the man is hostile, not kind. Or that it happened in the nineteenth century, not before. But there is only *one* universe. We cannot compare the universe with anything. We cannot *say* the universe.

I call the universe a green dragon because I want to avoid lulling you into thinking we can have the universe in our grasp, like a stray dog shut up in its kennel. I want to remind us of this proper relationship as we approach the Whole of Things.

On the other hand—and here is a second reason for the green dragon—we have learned things in our scientific explorations that completely transform our understanding of the universe. Our revolution in thinking dwarfs Copernicus's announcement that the Earth travels around the Sun. It is outrageous to compare the universe to a green dragon, I know, but I hope this will express some of my astonishment at what we now know about the universe. The inadequacy of the dragon image is that green dragons are much too commonplace to indicate the radical nature of what we have learned. That's how limited our language is.

So. Shall we begin?

YOUTH: You're going to tell me the story of the universe?

THOMAS: What better way to spend an afternoon in the gracious presence of the Hudson River? You must be ready for some confusion—much of what you hear will perplex you. Interrupt me whenever you wish to stop and reflect on something. Only by doing so will you come to hear the story in a significant manner; only then can you begin to feel the magnitude of what is breaking into human awareness.

YOUTH: Will this take a long time?

THOMAS: No, no. We'll be done before the Sun sets, and it's already over Hawaii. Pour yourself some apple juice. When things get difficult, remember this stupendous red oak; it's been here some four hundred years. Think of what it has lived through! Think of its patience, its perdurance, its continued vitality as it has learned to interact with all that came its way. And yet, here it is—its presence will help us through some of the rugged passages ahead.

YOUTH: Where should we start?

THOMAS: At the beginning. We need to start with the story of the universe as a whole. Our emergent cosmos is the fundamental context for all discussions of value, meaning, purpose, or ultimacy of any sort. To speak of the universe's origin is to bring to mind the great silent fire at the beginning of time.

Imagine that furnace out of which everything came forth. This was a fire that filled the universe—that *was* the universe. There was no place in the universe free from it. Every point of the cosmos was a point of this explosion of light. And all the particles of the universe churned in extremes of heat and pressure, all that we see about us, all that now exists was there at the beginning, in that great burning explosion of light.

YOUTH: How do we know about it?

THOMAS: We can see it! We can see the light from the primeval fireball. Or at least the light from its edge, for it burned for nearly a million years. We can see the dawn of the universe because the light from its edge reaches us only now, after traveling twenty billion years to get here.

YOUTH: We can see the actual light from the fireball?

THOMAS: When you see a candle's flame, you see the light from the candle. In that sense, we see the fireball. We are able to interact physically with photons from the beginning of time.

YOUTH: So we're in direct contact with the origin of the universe?

THOMAS: That's right.

YOUTH: I can't believe I didn't know this.

THOMAS: Scientists have only just learned to see the fireball. The light has always been there, but the ability to respond to it required a tremendous development of the human senses. Just as an artist learns to see a lakeshore's subtle shades and contours, the human race learns to develop its sensitivities to what is present. It took millions of years to develop, but humans can now interact with the cosmic radiation from the origin of the universe. We can now see the beginnings of time—a stupendous achievement.

YOUTH: It's amazing.

THOMAS: Most amazing is this realization that every thing that exists in the universe came from a common origin. The material of your body and the material of my body are intrinsically related because they emerged from and are caught up in a single energetic event. Our ancestry stretches back through the life forms and into the stars, back to the beginnings of the primeval fireball. This universe is a single multiform energetic unfolding of matter, mind, intelligence, and life. And all of this is new. None of the great figures of human history were aware of this. Not Plato, or Aristotle, or the Hebrew Prophets, or Confucius, or Thomas Aquinas, or Leibniz, or Newton, or any other world-maker. We are the first generation to live with an empirical view of the origin of the universe. We are the first humans to look into the night sky and see the birth of stars, the birth of galaxies, the birth of the cosmos as a whole. Our future as a species will be forged within this new story of the world.

YOUTH: But what about *my* future? What difference will it make for me?

THOMAS: To begin with, you will have to embrace your creative potential. The universe has unfolded to this point. It has poured into you the creative powers necessary for its further development. The journey of the cosmos depends on those creatures and elements existing now, you among them. For the unfolding of the universe, your creativity is as essential as the creativity inherent in the fireball.

YOUTH: How can I learn about my creativity?

THOMAS: Consider the creativity acting throughout the universe. Look there, and you will begin to understand how that same creative activity gathers you into its work as well.

The fireball was a cauldron of creativity. In it were created all the elementary particles of the universe. All that exists on Earth exists only through the elementary particles that emerged in that first epoch of the universe's unfolding.

After the fireball, stars and galaxies were created. We must realize that the creation of a galaxy is one stupendous activity. Could we manage that? Yet galaxies were created by the hundred billion, each with its hundred billion stars. And all of it dances, the stars swirling about each other, exploding, creating new stars, holding each other in the silence of the gravitational embrace. And these stupendously complex systems of being simply leapt into existence. When we reflect on the creativity inherent in the universe, we are overwhelmed both by its enormity and by its seeming effortlessness.

To learn about creativity, we must begin to understand the creativity of the Earth. We know of no other planet with Earth's creative power. I speak now of Earth as a whole, as a creative entity. Earth created the land masses, the mountain ranges, the atmosphere. The moon and Mercury created mountain ranges but their creativity ended long ago. Mars, too, created mountains and a thick crust and an atmosphere, but its significant creative evolution has ceased. The Earth, on the other hand, will continue to create for billions of years. Jupiter certainly created an atmosphere, but Jupiter will never be able to bring forth a continent; its great mass will remain gaseous far into the future. Only on Earth were the creative dynamics able to fashion such diversity, even on this elemental realm. Earth created the oceans, a stupendous feat. We have yet to find another ocean in this galaxy, another lake or river. We know of no others besides our own.

YOUTH: None?

THOMAS: We've found water vapor, and ice, but that is all. The creation of ice is a profound enough manifestation of creativity; there *was* no ice in the first billions of years of the universe. Or to have created water vapor, as did Venus, certainly reveals creative dynamics at work. But to have created the oceans and to have maintained them for four billion years is an accomplishment of which only Earth can boast. For all we know, there may be no other planet that has shown such creative power. An alarming thought, perhaps, but one that must be considered seriously until we have evidence indicating otherwise.

YOUTH: Oceans seem so ordinary.

THOMAS: Yes, they do, but that only reflects the ordinariness of our minds. When we take the whole universe as our fundamental frame of reference, we begin to appreciate the cosmic significance of running water. Only by establishing ourselves within the unfolding cosmos as a whole can we begin to discover the meaning and significance of ordinary things.

Earth was a cauldron of chemical and elemental creativity, fashioning ever more complex forms and combinations until life burst forth in the oceans and spread across the continents, covering the entire planet. This creativity advanced until flowers bloomed on every continent, then advanced further until the vision of the flowers and all beauty could be deeply felt and appreciated. We are the latest, the most recent, the youngest extravagance of this stupendously creative Earth.

YOUTH: Are we the last?

THOMAS: We haven't even begun! How can you speak of an ending? We are only just starting out on the human venture, and are only too keenly aware of our immaturity. Even this discussion reveals the way in which the human self-reflexive mind continues to unfold. Only minutes ago, you were unaware of the primeval fireball. The entire species was unaware of the fireball's light for millions of years. Do you see? The universe continues to unfold, continues to reveal itself to itself through human awareness.

YOUTH: When you talk about how the oceans emerged, I can see that as an obvious addition to the Earth. But what do humans add that is actually new?

THOMAS: The human provides the space in which the universe feels its stupendous beauty. Think of it this way: before the human arrived, the Earth and universe were magnificent realities. However, some of the depths of this magnificence were yet to be felt, yet to be appreciated. We enabled some of the depths of the universe to be tasted, and we have only just begun our venture; much waits on our maturity. Why else do all the things of the Earth keep calling to the humans, hoping to feel their existence in an intensely felt living of life? Think of it. Even yesterday you did not allow the fireball entrance into your awareness. Now that you have heard, aren't you filled with its wonder?

YOUTH: Well, yes—very.

THOMAS: The universe shivers with wonder in the depths of the human. Do you see? Think of what it would be like if there were no humans on the planet: the mountains and the primeval fireball would be magnificent, but the Earth would not feel any of this. Can you see the sadness of such a state? The incompleteness?

I sometimes think the primary deed of a parent is to see the beauty and grace of children. Children are magnificent, gorgeous beyond telling. They themselves have no idea of what beauty they embody. Can you see the tragedy of a child with no one to feel and cherish its beauty? No one to fall in love with this magnificent creature? No one to celebrate its splendor?

The cosmos is the same: humans can house the tremendous beauty of Earth, of life, of the universe. We can value it, feel its grandeur.

YOUTH: And you're saying we haven't finished this work?

THOMAS: Each of the three eras of humanity has achieved its own unique vision of beauty. During the tribal-shamanic age, the great mysteries of Earth and sky and Sun burst into human consciousness. Think of what that must have been like! To have lightning sear the human mind for the first time, thunderstorms fill human feeling! Whenever we are moved to awe by the branching fire of lightning, whenever we tremble with expectation in the predawn forest, we are remembering the Earth's first taste of its own beauty.

In the second era of human history, that of the great classical civilizations, we witness the rise of the Chinese, the Indian, the European, the Middle Eastern, the Amerindian. By enabling humans to specialize in their work, civilizations brought forth human powers unimagined in the tribal world. In this matrix the great scriptures of the world were written, the classical spiritual disciplines forged. During this period of human history, there developed an appreciation for the human world as the intersection of the transphenomenal and phenomenal realms.

The scientific-technological era is humanity's third phase of development. In these last few centuries we have empirically penetrated the dynamics governing the Earth and the cosmos. The gravitational, the electromagnetic, the weak and strong nuclear interactions were discovered and encoded in mathematical language. The power to alter the Earth's dynamics through technological inventions was seized. The immensity of time and space dawned within human awareness, and even the origins of the universe burst into the individual, self-reflexive awareness. The scientific-technological period has enabled the dynamics of the universe to unfold in human consciousness.

At present, the human species moves into its fourth era, what we might call the age of the Earth. This does not mean that science or technology will evaporate. The tribal-shamanic era did not disappear when classical religious civilizations emerged, nor did these civilizations all disappear when the scientific-technological era began. But the creative fire within the human venture now focuses on bringing forth something entirely new, a form of human life that envisions itself within the interconnected dynamics of the unfolding Earth reality. The tribe will not be the center of the human world, nor will the civilization, the culture, nor the nation-state. It will be the Earth community as a whole that will be understood as our home, our womb of creativity and life.

Humans will demand a deeper awareness of the planetary and cosmic dimensions implicit in self-reflexive mind. From the planet's point of view, we can say that the Earth is awakening to its own beauty, power, and future possibilities. The Earth awakens to the unfolding vision of a self-aware entity.

YOUTH: The Earth is a person?

THOMAS: No. The Earth awakens *through* the human mind. You have to understand this from two different points of view. We have a humanity that awakens to its planetary dimension, to its planetary responsibility, and thus begins to provide the Earth with a heart and mind. From the other perspective, we can see how the planet as a whole awakens through self-reflexive mind, which happens to unfurl through humanity.

YOUTH: Does everyone know about this?

THOMAS: The great spasms of confusion that torture so many of us in this time are manifestations of some degree of awareness of our situation. Despair and fear are ways in which many people reveal the repressed awareness that something of immense proportions is taking place on Earth.

YOUTH: You say we're leaving the scientific-technological era. What happens to science and technology then?

THOMAS: During the scientific-technological period, we regarded technology as a way of improving the human lot. We regarded science as knowledge that humanity had gathered concerning the universe. But during the age of the Earth, we will learn to see both science and technology as activities of the Earth. There were plant technologies on this planet for hundreds of millions of years before the emergence of humans, and similar scientific knowledge throughout the biological world. Or did you think that weather forecasting was a human invention? We will begin to understand that science and technology have emerged to serve the planet's unfolding, to enhance the total fabric of Earth's reality, and not merely to serve us. You see, humanity is a creation of the Earth process; we were brought forth to enrich the total life of the planet with our science and technology and everything else.

YOUTH: But what can I do? How am I supposed to help out?

THOMAS: Don't get impatient. You have to learn first. Just moments ago the presence of the universe's origin was unknown to you. Be patient, for there is certainly specific work waiting for you. Or did you think that the universe went to twenty billion years of work to create you if there was not a particular function that you—and *only* you—could do? The creative powers residing in you will be evoked in time for the work they were created for.

YOUTH: What creative powers?

THOMAS: We can not say until they show themselves. Not even you could know yet.

YOUTH: But where do they come from then, if even I don't know what they are?

THOMAS: From the same place that everything comes from. From the same place out of which the primeval fireball comes: an empty realm, a mysterious order of reality, a no-thing-ness that is simultaneously the ultimate source of *all* things.

YOUTH: Now wait a minute—

THOMAS: I realize how strange it sounds. But there is little we can do about that. I'm speaking here of something that has recently been encountered empirically. In the language of physics, we call it quantum fluctuation. Elementary particles fluctuate in and out of existence. What a strange realization! Don't think that physicists have any easier time of it than you! Elementary particles leap into existence, then disappear. A proton emerges suddenly—where did it come from? Who made it? How did it sneak into reality all of a sudden?

We say it simply leapt out of no-nothing-ness. There was no particle, then there was. I am not speaking here of the mannner in which mass and energy can be transformed into one another. I am speaking of something much more mysterious. I am saying that particles boil into existence out of sheer emptiness. That is simply the way the universe works. We have to get used to it. We didn't construct it; we just find ourselves here. If elementary particles are going to come leaping out of mysterious realms, then that's the way it is.

I say no-thing-ness. Or emptiness. But this only reveals the limits of language. We are here approaching an Ultimate Mystery, something that defeats our attempts to probe and investigate. There was no fireball, then the fireball erupted. The universe erupted, all that has existence erupted out of nothing, all of being erupted into shining existence.

What I would like you to understand is that this plenary emptiness permeates you. You are more fecund emptiness than you are created particles. We can see this by examining one of your atoms. If you take a single atom and make it as large as Yankee stadium, it would consist almost entirely of empty space. The center of the atom, the nucleus, would be smaller than a baseball sitting out in center field. The outer parts of the atom would be tiny gnats buzzing about at an altitude higher than any pop fly Babe Ruth ever hit. And between the baseball and the gnats? Nothingness. All empty. You are more emptiness than anything else. Indeed, if all the space were taken out of you, you would be a million times smaller than the smallest grain of sand.

But it's nice knowing we are this emptiness, for this emptiness is simultaneously the source of all being. You see?

YOUTH: And this too has just been discovered?

THOMAS: Yes. The way particles spontaneously leap into existence is a radical discovery in our own lifetimes. All of this is new within the scientific enterprise, breaking with traditions that go back to the beginning of science.

But from another point of view, we are arriving at an understanding that was deeply appreciated during the classical religious period of humanity. Thomas Aquinas and Meister Eckhart in the Middle Ages of Europe grasped intuitively that emptiness is the source of everything. They understood this realm of the not-put-together as the ultimate simplicity of the Godhead. This realization is echoed in the life and teaching of Buddha, who understood that all put-together things arise from emptiness and exist inseparably with emptiness.

YOUTH: Do physics and Christianity and Buddhism say the same thing?

THOMAS: Nothing that simplistic can be said. The situation is this. The creation story unfurling within the scientific enterprise provides the fundamental context, the fundamental arena of meaning, for all the peoples of the Earth. For the first time in human history, we can agree on the basic story of the galaxies, the stars, the planets, minerals, life forms, and human cultures. This story does not diminish the spiritual traditions of the classical or tribal periods of human history. Rather, the story provides the proper setting for the teachings of all traditions, showing the true magnitude of their central truths.

We are forging a cosmology that embraces humanity as a species, one that does not ignore the special cultural contributions of each continent, but that enhances these differences. Each tradition is irreplaceable. Not one can be reduced to any other. Each is vital to the work of the future. Each will flower beyond telling in fruitful interaction with the rest in the overall embracing story of the cosmos.

During the first centuries of the modern period, such a situation was impossible. An antagonism existed between modern ways of knowing and traditional ways of life and belief. Perhaps this was necessary; the scientific enterprise needed austere isolation from both the animistic attitudes of the tribal period and the spatial cosmologies of classical civilizations. Scientific understanding was too new and too different to fit into previously existing modes of human awareness; it needed to establish its own canons, procedures, and experiments without reference to anything outside itself.

The great wonder is that this empirical, rational journey of science should have any contact at all with spiritual traditions. But in our century, the mechanistic period of science opened out to include a science of mystery: the encounter with the ultimacy of no-thing-ness that is simultaneously a realm of generative potentiality; the dawning recognition that the universe and Earth can be considered as living entities; the awareness that the human person, rather than a separate unit within the world, is the culminating presence of a billion-year process; and the realization that, rather than having a universe filled with things, we are enveloped by a universe that is a single energetic event, a whole, a unified, multiform, and glorious outpouring of being.

We do not want to forget that the division between science and religion has created sufferings of all kinds. We have paid a tremendous price to establish scientific activity, and only by remembering the suffering that this schizophrenic situation created can we celebrate the present. We have a vast new empirically grounded story of the universe, one that explodes beyond any previous telling of reality, one that encompasses all peoples because it is rooted in concrete experience. Within this emerging story we can continue our journey to our fullest destiny.

YOUTH: What is our fullest destiny?

THOMAS: To become love in human form.

YOUTH: Love? I thought we were talking about science and religion. And emptiness.

THOMAS: Yes, that's right. The journey out of emptiness is the creation of love.

YOUTH: I'm confused.

THOMAS: By what exactly?

YOUTH: Well, by love. What do you mean by love?

ALLUREMENT

THOMAS: In order to approach love, we must start with our common context, the emerging universe in which we find ourselves. This realm of existence is our ultimate home. All beings, including humans, have this home in common. If we want to learn anything, we must start with the cosmos, the Earth, and life forms.

Love begins as allurement—as attraction. Think of the entire cosmos, all one hundred billion galaxies rushing through space: At this cosmic scale, the basic dynamism of the universe is the attraction each galaxy has for every other galaxy. Nothing in all science has been established and studied with greater attention and detail than this primary attraction of each part of the universe for every other part.

YOUTH: The attraction is love?

THOMAS: Start here: on the cosmic scale, an attraction exists.

YOUTH: But isn't that gravity?

THOMAS: Gravity is the word used by scientists and the rest of us in the modern era to *point* to this primary attraction. Listen carefully, and I'll clarify my point. For three hundred years, the word gravity meant Newton's theory of gravity. Then Einstein published his own relativistic theory of gravity, so that in our time a scientist can think of gravity as being Einstein's theory. The subtle mathematical differences between Newton's and Einstein's theories of gravity are crucial, but both were attempts to say something intelligent about why rocks fall to Earth. Before—and after—any theory, there is the ultimate mystery of the falling rock and the revolving Earth. The mystery remains no matter how intelligently we theorize. Do you see?

YOUTH: I don't think so.

THOMAS: All right. If a rock is dropped, why does it move toward the Earth?

YOUTH: Because of gravity.

THOMAS: And what is gravity?

YOUTH: A basic force. It pulls things.

THOMAS: *What is doing the pulling?*

YOUTH: There's just this pulling, that's all. It's just there.

THOMAS: That's right. An attracting activity. This attracting activity is a fundamental mystery.

YOUTH: But it's one we understand.

THOMAS: We understand details concerning the *consequences* of this attraction. We do not understand the attracting activity itself. Years after Isaac Newton wrote out his equations of the Universal Law of Gravitation, he was still wondering aloud: "Whence is it that the sun and planets gravitate toward one another?" We can never penetrate into the basic fact of this attraction, nor determine why it operates at all.

Do you see that the universe might just as well have been different? Might have included *no* attracting activity? But the fact is that our galaxy is attracted by every other galaxy in the universe; and our galaxy attracts every other galaxy. The attracting activity is a stupendous and mysterious fact of existence. Primal. We awake and discover that this alluring activity is *the* basic reality of the macrocosmic universe.

YOUTH: Are you saying that this attraction is love?

THOMAS: The difficulty with the word "love" is that it has been ruined recently. For the last few centuries the fundamental referent for our language was the human world. We have tried to live in the anthropocentric frame of reference and have thereby crippled many of our concepts and words. When we hear the word love we think only of *human* love, a very special sort of love. So I am certainly not saying that gravity is human love.

I *am* saying that if we are going to think about love in its cosmic dimension, we must start with the universe as a whole. We must begin with the attraction that permeates the entire macrostructure. I'm speaking precisely of the basic binding energy found everywhere in reality. I'm speaking of the primary allurement that all galaxies experience for all other galaxies.

YOUTH: How does this connect with human love, then?

THOMAS: Tell me something you enjoy doing.

YOUTH: Listening to music.

THOMAS: Yes. Now watch. We can not give any explanation for liking music; we simply enjoy music of certain sorts. The attraction is primal. You have awakened to existence and discovered this attraction. Is it clear now that your attraction, your interest and enjoyment, are ultimate mystery?

YOUTH: I'm beginning to see.

THOMAS: There are so many sounds in the world, and yet a very particular sort of sound interests you most deeply. Why should this be? Why not any of the other infinite number of sounds? Why music above all? Well, that is unanswerable, just as Newton never pretended to be able to say why the Sun attracts the Earth. The strangest thing is that this alluring activity permeates the cosmos on *all* levels of being. These allurements permeating you and everyone and everything else are fundamentally mysterious. You are interested in certain things, certain people, certain activities: each interest is as fundamental to the universe as is the gravitational attraction our Earth feels for the Sun. We cannot explain why these attractions exist. We can only become aware of them. Am I making myself clear?

YOUTH: Yes, but it seems that maybe we *can* explain them. For instance, listening to music is relaxing. Maybe that's why humans—

THOMAS: When you first listened to some music you really liked, did you think, "This is the sort of music that will relax me?"

YOUTH: Well, no.

THOMAS: You discovered that you were drawn to the music, true? Such experiences of interest are the roots of love. You are simply attracted to something or someone, to some activity. You don't find reasons for this attraction until after the fact; *then* you come up with reasons. The Earth does not think: "Well, it'll be a good thing to be attracted to the Sun. That way, humans can warm their tea in black bags and save on electricity." The Earth is simply attracted. The electron is simply attracted. The galaxy is simply attracted. *You* are simply attracted. This mysterious attraction that we call "interest," or "fascination," is as mysterious, as basic, as the allurement that we call gravitation.

YOUTH: So what you're saying is, a galaxy exists within attraction and so do I.

THOMAS: The great mystery is that we are interested in anything whatsoever. Think of your friends, how you first met them, how interesting they appeared to you. Why should anyone in the whole world interest us at all? Why don't we experience everyone as utter, unendurable bores? Why isn't the cosmos made that way? Why don't we suffer intolerable boredom with every person, forest, symphony, and seashore in existence? The great surprise is the discovery that something or someone *is* interesting. Love begins there. Love begins when we discover interest. To be interested is to fall in love. To become fascinated is to step into a wild love affair on any level of life.

Then we discover not only that we are interested, but that our interests are entirely our own. We awake to our own unique sets of attractions. So do oxygen atoms. So do protons. The proton is attracted only to certain particles. On an infinitely more complex level, the same holds true for humans: Each person discovers a field of allurements, the totality of which bears the unique stamp of that person's personality. Destiny unfolds in the pursuit of individual fascinations and interests.

YOUTH: But it almost sounds self-centered. Where do others fit in?

THOMAS: By pursuing your allurements, you help bind the universe together. The unity of the world rests on the pursuit of passion. Surprised? Let's experiment:

Bring to mind all the allurements filling the universe, of whatever complexity or order: the allurement we call gravitation, that of electromagnetic interactions, chemical attractors, allurements in the biological and human worlds. Here's the question: If we could snap our fingers and make these allurements—which we can't see or taste or hear anyway—disappear from the universe, what would happen?

To begin with, the galaxies would break apart. The stars of the Milky Way would soar off in all directions, since they would no longer hold each other in the galactic dance. Their spiralling arms would disintegrate as stars made their chaotic ways into intergalactic space. Individual stars would disperse as well, their atoms no longer attracting each other but wandering off in all directions, releasing core pressure and thereby shutting down fusion reactions. The stars would go dark.

The Earth would break apart as well, all the minerals and chemical compounds dissolving, mountains evaporating like huge dark clouds under the noon sun. And even if the physical world retained its shape, the human world would disintegrate just the same. No one would go to work in the morning. Why should they? There would be no attraction for the work, no matter what it was. Activity would cease. Did scientists once find the universe interesting, staying up nights to reflect on its mysteries? No longer. Did lovers chase each other in the night, abandoning all for the adventure of romance? Never again. All interest, enchantment, fascination, mystery, and wonder would fall away, and with their absence all human groups would lose their binding energy. Galaxies, human families, atoms, ecosystems, all disintegrating immediately as the allurement pervading the universe is shut off. Nothing left. No community of any sort. Just nothing.

YOUTH: That's an amazing experiment.

THOMAS: It underlines the primary result of all allurement, which is the evocation of being, the creation of community. All communities of being are created in response to a prior mysterious alluring activity. OK? Allurement evokes being and life. That's what allurement *is*. Now you can understand what love means: love is a word that points to this alluring activity in the cosmos. This primal dynamism awakens the communities of atoms, galaxies, stars, families, nations, persons, ecosystems, oceans, and stellar systems. Love ignites being.

Think of the power of this alluring activity—its immensity. We are barely able to keep our *cars* puttering about the continent! What would we say if we had the job of getting the stars to rotate and revolve around the galaxies? What if we had to keep all the hydrogen atoms together? Or keep them pressed into stars? Think of the tremendous galactic tasks performed every instant by this universe, and you will begin to feel the magnificence of the cosmic allurement of love. It is this allurement that excites lovers into chasing each other through the night, that pulls the parent out of bed for the third time to comfort a sick child, that draws humans into lifetimes of learning and developing. The excitement in our hand as it tears open a letter from a friend is the same dynamism that spins our vast Earth through the black night and into the rosy colors of dawn.

YOUTH: So this alluring activity is love?

THOMAS: Yes: the activity of allurement, which is simultaneously the activity of igniting being and enhancing life.

YOUTH: Is alluring the same as evoking?

THOMAS: Consider the star. The star's development indicates clearly how allurement and evocation of being are a single dynamism.

Imagine a vast dark cloud of hydrogen atoms stretching through millions of miles of space. Each of these trillions upon trillions of atoms is involved in an attracting activity for all the rest, and slowly begins to move. A common center emerges, and the hydrogen atoms begin to clump together. The growing pressure from the gravitational attraction enables the hydrogen atoms to fuse into helium atoms, thus releasing their hidden energy in a vast profusion of light emanating in all directions: the core of the star ignites. All of this activity is the result of the cosmic allurement of gravitation. First we had a black cloud of hydrogen atoms; now we have a stellar brilliance radiating through intergalactic space to the farthest reaches of the cosmos. We had only hydrogen; now we have the star. You see? The allurement of gravitation evoked the being of the star. The hydrogen atoms responded to this allurement and showed their deeper potentiality as elements of a raging star. Only by responding to allurement could the depths be shown, and the being of the star come forth.

YOUTH: And the same holds for humans?

THOMAS: The same holds for you, yes. You do not know what you can do, or who you are in your fullest significance, or what powers are hiding within you. All exists in the emptiness of your potentiality, a realm that cannot be seen or tasted or touched. How will you bring these powers forth? How will you awaken your creativity? By responding to the allurements that beckon to you, by following your passions and interests. Alluring activity draws you into being, just as it drew the star into being. Our life and powers come forth through our response to allurement.

YOUTH: No matter what allurement?

THOMAS: That's right.

YOUTH: How about reading Shakespeare? What would that pursuit evoke?

THOMAS: If you read deeply and are drawn into the dramas, you will ignite previously unsuspected capacities for being. You will evoke a spaciousness where the feelings of the human world can live. Plunge into the life of the plays, and one day you will be startled by the discovery of feelings you had not known before: an affection for the human condition, for the frailties of the human will, for the nobility of spirit that wells up in every generation, no matter how difficult the circumstances of suffering and disillusionment.

You asked about the others, how they fit in. Do you see now? A journey into Shakespeare's works enables you to enter more fully the complex relationships within the human world. You may abide more deeply in these relationships precisely because the ontological space within yourself has been opened up by the power of Shakespeare's language. You will enter relationships in a subtler manner, because your awareness has been opened. More of the world will be present to you; what was formerly invisible now shows itself. That's what we mean by saying your being has been aroused, activated, awakened, and evoked.

Pursue these interests further and you will learn what guided English society, ancient Roman society, and medieval Italian society. Led into an understanding of the way in which the past lives in our present world, you will begin to see how contemporary patterns of activities are shaped by the history of the West. You will carry within yourself the complexity of the world in a manner unimaginable to your previous self. You will know that you are not disconnected from the life of the world, nor from struggling humanity in all its difficulties throughout the planet. You will learn the first glimmer of the profound manner in which humans bind together the entire social order through a heightened awareness of what it means to be a compassionate human.

YOUTH: And that's what you mean when you say that what was invisible is made visible? I mean, all these subtle possibilities for relationships are suddenly noticed, are suddenly made present. You know, it's amazing to think what sort of world this would be if Shakespeare or other poets had not written. But why do they write? Is that allurement, or is that something else?

Our Destiny As Enchantment

THOMAS: Your question plunges us into the root of all mystery. We awake to a universe permeated with allurement, and our most primal desire is to become this allurement. We awake to a universe filled with fascination, and our most fundamental urge is to *become* this fascination.

YOUTH: I don't understand.

THOMAS: Stay with Shakespeare. You are drawn to the works of Shakespeare, say. Through these works you deepen your sense of community, making the ancient Romans intimately present to you in a new way. Because of these newly evoked creative powers of perception, you enter more effectively into intimate relationship with the people of your own time and place. You appreciate feelings that others might have, and intuit their motivations. Thus you enter into more complex relationships within human groups. All of this from reading and studying Shakespeare. He wrote his plays, and through them you enter more deeply into being.

So do you see that Shakespeare is inseparably involved with allurement? Do you see that Shakespeare's works are evocations of being?

YOUTH: I'm still somewhat confused...

THOMAS: The question you asked was: Why did Shakespeare write? He wrote because the world enchanted him. He wrote to capture the grandeur, pathos, profundity, and beauty that he experienced in life. In order to do so, he had to become one with this beauty. How else can we express feelings but by entering deeply into them? How can we capture the mystery of anguish unless we become one with anguish? Shakespeare lived his life, stunned by its majesty, and in his writing attempted to seize what he felt, to capture this passion in symbolic form. Lured into the intensity of living, he re-presented this intensity in language. And why? Because beauty stunned him. Because the soul can not confine such feelings.

Shakespeare put himself into writing because by writing he could fascinate others, just as the world had fascinated him. He could amuse, astonish, delight, and enchant others just as the world had enchanted him. Drawn into life by allurement in a thousand different ways, he himself then became alluring. Stunned by the fascination permeating the human order of existence, he in turn fascinated.

YOUTH: This is true for poets most of all, but . . .

THOMAS: No, no, not at all. Consider scientists—Stephen Hawking comes to mind. Here is an astrophysicist fascinated by the primeval fireball, the initial singularity of space-time. He pursued this path further into experiences of order and beauty, of the complexity and simplicity of the universe's earliest moments. So what does he do? He articulates his experience within the languages of English and mathematics. He creates his own magnificent language forms to communicate the beauty he has uncovered, the clarity he has achieved, the insight he has seized. He hopes to capture some of this, luring others into similar moments of seeing, captivating their minds, drawing them more deeply into their own understanding and feeling of the universe. The beauty of his mathematical language is as alluring as Shakespeare's iambic pentameter. Mathematical physicists can not resist the lure of Hawking's creations; they seize the mind as powerfully as Shakespeare's. Do you see what I am saying?

YOUTH: That we awake to fascination in some form or another, and that we strive to become fascination.

THOMAS: Yes, we awake to fascination and we strive to *fascinate*. We work to enchant others. We work to ignite life, to evoke presence, to enhance the unfolding of being. All of this is the actuality of love. We strive to fascinate so that we can bring forth what might otherwise disappear. But that is exactly what love does: Love *is* the activity of evoking being, of enhancing life.

YOUTH: Now, this is human love you are describing?

THOMAS: No, no, no. You must begin to see this activity as basic to the universe. Consider the star again. In the core of a star helium, carbon, oxygen, silicon, all the elements up to iron are created in blazing heat. If a star is of sufficient size, after billions of years it explodes, creating all the rest of the elements, sending them off into the universe. Our own solar system emerged from an exploded supernova, creating the planets and their many elements. Minerals and life forms are created out of supernova explosions.

Think about it! When you breathe, you breathe the creations of a star. All the life you will live is possible because of the gifts of that star. Your life has been evoked through the work of the heavens, do you see? The star emerges out of its own response to allurement, then evokes the life of others. The air we breathe, the food we eat, the compounds out of which we are composed: all creations of the supernova.

Drawn into existence by allurement, giving birth, then drawing others into existence—this is the fundamental dynamism of the cosmos. In this we can see the meaning of human life and human work. The star's own adventure captures the whole story. It is created out of the creations of the fireball, enters into its own intense creativity, and sends forth its works throughout the galaxy, enabling new orders of existence to emerge. It gives utterly everything to its task—after its stupendous creativity, its life as a star is over in one vast explosion. But—through the bestowal of its gifts—elephants, rivers, eagles, ice jams, root beer floats, zebras, Elizabethan dramas, and the whole living Earth, become possible. Love's dynamism is carved into the principal being of the night sky.

YOUTH: Are you saying that the star is aware of what it is doing?

THOMAS: Well, yes and no. But let's think about it a moment. We are the self-reflexion of the universe. We allow the universe to know and feel itself. So the universe is aware of itself through self-reflexive mind, which unfurls in the human. We were brought forth so that these experiences of beauty could enter awareness. The primeval fireball existed for twenty billion years without self-awareness. The creative work of the supernovas existed for billions of years without self-reflexive awareness.

That star could not, by itself, become aware of its own beauty or sacrifice. But the star can, through us, reflect back on itself. In a sense, you are the star. Look at your hand—do you claim it as your own? Every element was forged in temperatures a million times hotter than molten rock, each atom fashioned in the blazing heat of the star. Your eyes, your brain, your bones, all of you is composed of the star's creations. You *are* that star, brought into a form of life that enables life to reflect on itself. So, yes: the star *does* know of its great work, of its surrender to allurement, of its stupendous contribution to life, but only through its further articulation—you.

YOUTH: So that star is only just now aware of its work?

THOMAS: Yes, in the same way that you become aware of aspects of yourself that have remained unconscious for years. You've seen pictures of yourself as an infant; when you look at them, you're looking at yourself. The baby then becomes aware of its own beauty. Aren't you the further development of that baby? Well, of course, and yet the baby seems somehow other than you.

So, too, with the star. We know we are the further development of the star, and yet we know that we are somehow different from the star. The star emerges into self-reflexive awareness of its beauty and creative work through the human mind.

The universe is a single multiform event. There is no such thing as a disconnected thing. Each thing emerged from the primeval fireball, and nothing can remove the primordial link this establishes with every other thing in the universe, no matter how distant. You and everything you do and become are further articulations of the primal fireball.

Humans have always been fascinated by genealogical records. We want to know where we came from, the history that leads up to *us*. But nothing in all genealogical research throughout human history could have prepared us for the truth. Several hundred years or several thousand years of a family tree is nothing, for our common family tree fills the universe. Our kins include the living companions on Earth, all the planets and stars, and every galaxy. Cousins all, we spread out until we fill the whole vast cosmos, end to end.

The people of the Middle Ages had the right idea in their cherishing of relics. If they could find a splinter of the cross, or a garment worn by St. Francis, they venerated the article because it had once been so close to beings of extraordinary significance. This attitude must be extended. Our reverence for the holy must expand to include the whole numinous universe. What are the relics today? *We* are the relics, the Earth and all beings of Earth were there in the core of that exploding supernova. We were there in the distant, terrifying furnace of the primeval fireball. Not as mere witnesses, either, but as central to the event. Our bodies remember that event, exulting in the majesty of the night sky precisely because all suffered it together. The planet is a rare and holy relic of every event of twenty billion years of cosmic development.

When we deepen our awareness of the simple truth that we are here through the creativity of the stars, we begin to feel fresh gratitude. When we reflect on the labor required for our life, reverence naturally wells up within us. Then, in the deepest regions of our hearts, we begin to embrace our own creativity. What we bestow on the world allows others to live in joy. Such a stupendous mystery. . .!

Think of it. This supreme dynamic of love, of allurement and evocation, in action since the beginning of the universe, after billions of years becomes aware of itself. Life-enhancing and being-evoking allurement knows itself, the magic of creating life and being now reflects upon its own mystery! What creatures, what living beings, what persons will follow us, entering life and the great mystery of love precisely because of our work?

Let's speak here of values. I don't mean the values of modern society or philosophers, or the values of the market place; I mean cosmic value. What does the cosmos value? What does the cosmos, as the supreme home of all, itself value? Those who awake to the splendor of the universe and ignite life in others.

Tell me. Are you aware that you and you alone are able to draw forth life in ways that no one else in the universe can?

YOUTH: You mentioned Shakespeare and that astrophysicist. . .

THOMAS: My question has nothing to do with them! The universe would never bother to create two Shakespeares. That would only reveal limited creativity. The Ultimate Mystery from which all beings emerge prefers Ultimate Extravagance, each being glistening with freshness, ontologically unique, never to be repeated. Each being is required. None can be eliminated or ignored, for not one is redundant.

Are you aware of the ways in which you have the power to evoke being? This question probes your destiny as a creative source, your ultimate value. To answer requires that you move more deeply into the primordial dynamism of the universe, for as you ripen into love's activity you simultaneously enhance the life around you.

YOUTH: I don't know where to begin thinking about it.

THOMAS: Begin with your allurements and your own web of relationships. Your allurements draw you into the activity of evoking the life about you. There are persons and creatures all around you that will emerge into an enhanced vitality, with a renewed taste for life's adventures, only if you pursue your destiny with the same extravagant devotion of the star to its destiny.

YOUTH: I'm to become like a star?

THOMAS: In its pursuit of allurement, yes. In its complete immersion in the work at hand, in its identification with the activities of arousing being, yes. There are so many beings you can emulate: the simplest prokaryotic organisms struggled ceaselessly and with stunning success, altering the nature of the Earth permanently. They roamed through life and hatched those seeds of power we call genes. Who could have created them if they had not? We have no talent for that kind of work. We carry their achievements in our bodies. All the hundreds of thousands of genes in our bodies that enable such lambent beauty to delight the planet were handed to us by these primitive creatures.

Your gratitude includes them. Your life emerges through their creativity.

YOUTH: But they didn't know what they were doing. I don't see how I can be grateful to them for their mindless behavior.

THOMAS: Do you know what *you* are doing?

YOUTH: More than they.

THOMAS: I would hope so, yes. Unless their labor was in vain. But do you know what you are doing when you find Shakespeare so fascinating? Do you know what's happening, in a cosmic sense? Can you explain to me quite simply why humans find mountains magnificent beyond capture in language, why they risk their lives to be up there on the angular planes of granite?

YOUTH: Well, no. Not in any ultimate sense.

THOMAS: Then you share the same cosmic ignorance with the microorganisms who created the informed sequences of nucleotides we call genes. Neither you nor they understand why the cosmos should glimmer with beauty, drawing forth our deepest efforts. The simple truth is that we do pursue the fascinating beauty that surrounds us. Can you tell me what will come of your creativity and your destiny? Of course not! Nor could the microorganisms predict the future or speak of the meaning of their labor in any ultimate sense. We are similar in hoping to immerse ourselves in the life-evoking activities that fill the Earth.

We pursue being and struggle to become one with this enchanting mystery, so that we too can contribute to life just as they did. We suffer the withering pains of life in the hope that we, too, just as the stars and the prokaryotes, can enter the adventure of the cosmos and enhance the riches of the universe.

YOUTH: Then how can I learn to become love?

THOMAS: That is the easiest thing to do in the universe! All that is required is that you fall in love. Fall in love as deeply as you can. That way the universe becomes your primary teacher. We learn about becoming love by falling in love, then reflecting on the experience and what it taught us. That way, the universe does the real teaching. I'm not talking about second-hand, theoretical, or abstract ideas *about* love, but immersion *in* the actuality of love and love's activity. Remember that the desire to make us over into love permeates the universe. We are initiated into love when lured into the intense pursuit of the enchanted lover. If the initiation is long, and filled with doubt and suffering, the learning takes hold deeply.

YOUTH: Why is that?

THOMAS: The *slow learner* has so many more opportunities to watch the dynamics of love's play. If he is the most stubborn human in the history of the Earth, he will have had the opportunity to see how shrewd love can be as love penetrates all his character armor. When the stubborn human finally falls in love, he will understand how hard the universe had to labor to finally win him. He will know something of the subtle arts of love: how protean, how untiring, how confident, how intelligent, how faithful, how boundless, how enflamed, how unitive, how irresistible love can be. Such stubbon humans become the world's greatest lovers, for they have been through an initiation that demanded many resources of love; they make themselves just as irresistible and as intelligent as love in drawing others into the joy of living.

YOUTH: But this is so idealistic. I mean, fine, I like it, but I know what my father would say: It has nothing to do with the real world of business and everything else. He's an accountant. How does any of this apply to an accountant?

THOMAS: During the last few centuries of the modern period, we tried to use the human as the locus of all meaning. By so doing we have ruined our language. We can see this in the very word "accountant." We think of an accountant as someone who keeps the books of a company, who keeps track of sales and inventory, watching over the bottom line of profits. Within that small world it is difficult, if not impossible, to sense the involvement of self, Earth, and cosmos. So in order to answer your question concerning the dynamic of love within the profession of accountancy, we need to tell the full story of the cosmos.

To begin with, the Earth is a corporation. It is the primary corporation. Any corporation created by humans must fit itself into the larger corporation of the Earth, because if the Earth goes bankrupt everything else falls to ruin. Furthermore, the Earth has its own system of accounting, much more subtle and stringent than the human system of profit-and-loss. The Earth keeps track of all energy exchanges, no matter how minute. The Earth's books remember all materials used and all waste created in production. There is no carpet under which anything can be swept: All is counted and entered in the natural ledgers of the Earth. Now the question is this: What is an accountant in a human corporation within this larger context?

Suppose a firm manufactures shoes of various types. An accountant would begin by realizing that everything required for production is provided by the Earth. The leather is created by the animals, the dye is created by the mineral world, the Sun and plants provide energy, and humans provide the intelligence that knits everything together. Even the motivations are provided by the Earth: craftspeople desire to exercise their skills, and humans in general deeply desire to be of use, to enter a community of abiding mutual relationships.

Accountants would see themselves as essential to the whole process. Creating procedures that enable the gifts of the animal, mineral, and human worlds to present themselves, they enhance the life process. If the corporation is well-ordered, everyone's joy will blossom, for people purchasing the shoes will know the simple joy of wearing shoes of true quality, and those making the shoes will know the deep joy of living a life of worth. In the corporations of the future, accountants will know a new joy when they realize that their work enables the life of the entire bioregion to prosper. The overall vitality of the particular geographical region of the Earth will be enhanced, rather than degraded, by the corporation's interactions with the sea, sunlight, air, life forms, and soil. Then the joy of the fish and grasses and of the soil communities will be added to the joy of humans, and all of this will be the way in which an accountant, or any corporation executive, enters into the dynamics of love.

YOUTH: I see. This is a different way of...

THOMAS: When you take the story of the universe as your basic referent, all of your thoughts and actions are different.

YOUTH: So it's not just accountants.

THOMAS: All professions, all work, all activity in the human world finds its essential meaning in the context of the cosmic story.

YOUTH: And everything's so fouled up now. Here we are on the verge of blowing up the Earth. Why is it so bad? Why are we so violent? Why can't we just avoid all this suffering that we see everywhere? Are people ignorant of all this stuff you're talking about? Or is it something else?

Evil From Cosmic Risk

THOMAS: To begin with, understand that humans are not unique in having to suffer. Nor are humans unique in being violent. We live in a violent universe. Violence fills the cosmos in various forms, and human violence is only one of these. Violence is a universal fact, but not the dominant fact of the universe. The great mystery is not violence, but beauty. We note the violence, all the more amazed that such stupendous graciousness and beauty should exist anywhere at all.

YOUTH: But where does violence come from?

THOMAS: Destruction has its root in the allurement permeating the universe. Allurement is the source of all activity, even destructive activity. The star, responding to allurement, destroys itself. No one comes from the outside to demolish the star. The star implodes, smashing itself into a trillion parts—its journey is ended. Or imagine the violence of two stars colliding under mutual gravitational attraction. The fire would be splashed in every direction for millions of miles. Such tremendous violence, yet see the graciousness of hundreds of billions of stars swirling in the galactic dance.

The biological world knows all sorts of violence. The same urge that draws the lion to the river for water draws it on to kill the wildebeest. Insects are so intent to stretch forth and explore the world that they will devour their own parents if they cannot find other food. Fascination with living, the enchantment of being alive, the beauty of the surrounding world—all these draw creatures into violent acts and into the destruction of being, but after four billion years of life on Earth, what beauty has blossomed forth! There is danger in the natural world, a constant challenge, excitement, violence, risk, and terror, but out of this emerges the wonder of the Earth.

With the human a new quality of violence enters the Earth system, one coming from the power of self-reflexion. This new awareness is a risk as well as an achievement of the life process. In a sense, the earth wounded itself when it took on self-reflexive sentience: there appeared new powers of creativity, new dangers of destruction. The question hanging in the solar system today is this: Will the Earth benefit in beauty by risking human self-reflexive awareness? Or will the Earth suffer new and permanently crippling violence?

We see the singular beauty that has been achieved through violence in the cosmic and earthly realms. We do not know yet if the same will be the case in the human. Indeed, through the millenia of civilization, humans have seldom stopped and reflected seriously on whether or not we are beneficial additions to Earth's system of life. Self-absorbed, we focus on our own survival and on explorations into our innate powers. We never developed a larger perspective by which to evaluate our activities, a perspective that included stars, planets, and all other life forms. This limited world-picture is precisely what is ruining us as a species.

If we want a larger viewpoint, we need to examine the history of the Earth over the last ten million years say. The first human types entered the Earth's life perhaps three and a half million years ago, though some scientists push this date back even further. What we know now is this: Over the last ten million years, though many species have gone extinct, an even greater number of species have been created. The Earth's natural fecundity adds ever greater diversity to the total abundance and variety of life.

This situation of an ever-renewing life system has been reversed with the arrival of technological humanity. We have multiplied extinction rates many times over. The best estimates now show that the Earth loses a species every twenty minutes. We will lose at least half a million species in the next fifteen years. No one pretends to be able to predict what this will mean for the overall vitality of the Earth system, but one conclusion is inescapable: in our anthropocentric myopia, we humans are mauling the Earth's life forms. A thermonuclear war would only finish the global wake of chemical toxic destruction already well under way on every continent.

Can Earth sustain our violence? Can a great beauty grow from the ruins we leave? Concerning this question, it is important to understand the temporal nature of the Earth's creativity. The Earth at one time was able to create life, but that time has gone. The first life forms consumed the very conditions that enabled life to emerge. The fertility of the Earth is different now. If the higher life forms disappear, they can not be re-created. When life forms vanish, they vanish forever. The situation is similar to a young child raised outside human language communities: after its first few years, the child will *never* be able to develop language. The neurophysiological connections needed to carry the language functions are present only during the first few years, then disappear. If language is not developed then, it can not be developed in the future.

YOUTH: But the Earth would be able to do *something*, wouldn't it?

THOMAS: The Earth will continue on some level, no matter *what* humans do. But if we continue our chemical and nuclear assault on the planet, all future possibilities will be severely limited. To expect Rembrandt to create a new painting is fine, but if you first remove an eye and large portions of his brain you will have to accept what he is able to give you out of his diminished capacity.

We are soaking all life forms with poisons, changing rivers into lethal sewage, and hurling millions of tons of noxious gases into the respiratory system of the Earth. As scientific as we claim to be, we have yet to realize that babies do not come from storks. The simplest, most empirical fact is that babies of every species are created out of soil, air, rain, food, and rivers. If we change all of these into poison, we must accept the fact that we change our unborn into poison as well. What materials will be used for their arms but the minerals of the poisoned continents? Of what stuff will their eyes be fashioned but the water of our lethal rivers? What will those wet fleshy brains be made of but noxious gases and acid rain? Serious birth defects in the human world alone have already doubled in the last two decades.

To begin to evaluate the achievement of the humans, we might take a democratic vote. Let's not be chauvinistic here—let *everyone* vote. There are ten million species presently alive on the planet. Convene the United Species Conference, giving each species one vote, and put this question to the test: "Should the human species be allowed to remain within the Earth's system of life?" Imagine the debate. Our single representative would attempt to persuade 9,999,999 others that the human species is indeed worth keeping. Perhaps our representative will mention poetry. Perhaps religious or scientific or artistic creations. Now imagine the other species seated around the great table, weighing these contributions against all the Earth-killing poisons humans have planted in every continent, sunk into every ocean, launched into the sky.

YOUTH: But why? Why was there such a jump in violence with us? Why couldn't we blend in the way other species blended in?

THOMAS: This is the danger of self-reflexive awareness, what I mean when I say Earth in a sense wounded itself by allowing self-reflexion to emerge. The human is dangerous precisely because the universe is sublime. Here is the real question: "Can the cosmos survive the vision of its own beauty?" Can the Earth continue to create beauty once it has created a mirror to this beauty? Can the Earth continue to organize its unfolding once its depths of eros have been tasted, their sweetness enjoyed?

Humans reach into erotic intensities that are present everywhere in the natural world, but with the crucial difference of self-reflexion. The situation is clear in sexuality. Animals enter the delights of sexual intimacy only during female estrus. For the red fox, this lasts less than a single February week. For the palolo worms the period of mating is confined to a specific time of a single day of the year. Humans, however, can devote an entire life to the pursuit of sexual delight. The delight is felt within an awareness that it is felt. And that is the risk that the Earth took.

Why? So that the hidden reaches of the Earth process could be probed and felt and savored. More than the great whales, humans are creatures of the depths: the depths of all things. We are the space in which the universe can be cherished in a new and intense manner. So the question remains: Can this voluptuousness be contained within the human vessel? Can allurement bear the knowledge of its own essence? Or will the tensions this creates shatter any self?

YOUTH: You're saying that beauty and allurement are at the root of all evil activity?

THOMAS: Yes.

YOUTH: But what about nuclear weapons and the possibility of nuclear war? How is that the result . . .

THOMAS: Nuclear weapons would not exist if scientists and technologists had not been fascinated by the cosmos and overwhelmed by the possibility of penetrating into the deepest realms of reality. The idea of tapping the awesome sources of power is irresistible for the human mind. These fascinations are the root of our devotion to creating thermonuclear devices. It is not that alone, of course. Our political convictions too are the result of allurement. Soviet citizens are drawn to the dream of a worker state and a classless society, just as American citizens are drawn to the dream of a free enterprise system where everyone has plenty. Both nations are entranced by these visions, drawn to them, and out of these commitments build such grotesque weapons.

YOUTH: Then what goes wrong?

THOMAS: Humans are easily addicted to beauty, even a clouded vision of it, and we can not break the addiction. Our agricultural processes poison our water and destroy four billion tons of topsoil on the American continent each year, and still we keep at it. We are captivated by our consumer lives, addicted, and apparently nothing can break through. Unable to see the simple sadness of our way of life, sunk into our addictions, we overstuff our homes and garages, carrying on, unmoved by the smoke rising over the burnt-out lives of fifty other nations and a million other species. The American mind resembles a glove compartment, jammed tight with useless junk that no one pays any attention to until we consider cleaning it out; and even then, even as we wonder why we so needlessly clog up our lives, unable to part with it all, we just jam it back in its place.

The way to break an addiction is to break out of a limited world view. Break out of egocentricity. Break out of ethnocentricity. Break out of anthropocentricity. Take the viewpoint of the Earth as a whole. In every fascination, in every allurement, include the vitality of the Earth. You are the Earth too. The Earth is not different from you. This planet bloomed through millions of years and arrived at the stupendous achievement of self-reflexion. She surpassed herself, shivering with joy at the thought of housing a creature through whom her depths, her beauty, her majesty could be cherished in a new intensity. Imagine Earth's astonishment to see us attempt to satisfy ourselves by transforming the Earth into throw-away tinsel, most of it noxious to all forms of life. Imagine the hilarity and pathology of a civilization devoted to stacking up this stuff, instead of plunging into the joy that has been prepared over billions of years.

YOUTH: Then why didn't the Earth bring forth humans who were born free of our liability? You say our minds fixate on partial visions, that we forget the whole, the Earth, that we become addicted. Why didn't the Earth avoid all the destruction we inflict?

THOMAS: Our task is to explore, to celebrate and delight in the depths of the universe. To enter this work often involves tremendous suffering. You ask, "Why can't we be excused from our destiny?" We can be excused from this task only if some other species accomplishes it for us. Does that option appeal to you? To have something else do the work of the human? To suddenly have no worth or value whatsoever for the whole? In that case, why would the universe bother with us at all? We would have nothing to contribute. We would be, at best, only troublesome stowaways on the great cosmic journey.

YOUTH: Then tell me how I can know the difference between an allurement that will lead to beauty and one that will not? I mean, suppose I am not interested in stacking up consumer objects, or stocks and bonds and bank accounts. . . .

THOMAS: There is no rule that can be captured in language and applied from outside of particular situations. Reality is too complex, too subtle, too mysterious to submit to our demands to control it in this or any other way. The realization that one responds to the depths is as subtle an achievement as the ability to respond to light from the primeval fireball.

There are some central articulations that can serve us in our reflections. You can examine your own self and your own life with this question: Do I desire to have this pleasure? Or rather, do I desire to *become pleasure*? The demand to "have," to possess, always reveals an element of immaturity. To keep, to hold, to control, to own; all of this is fundamentally a delusion, for our own truest desire is *to be* and *to live.* We have ripened and matured when we realize that our deepest desire in erotic attractions is to become pleasure with and for our lover, to enter ecstatically into pleasure so that giving and receiving pleasure become one simple activity. Our most mature hope is to become pleasure's source and pleasure's home simultaneouly. So it is with all the allurements of life: we become beauty to ignite the beauty of others.

The history of life can be understood as the creation of ever more sensitive creatures in a universe where there is always another dimension of beauty to be felt and savored. Think of yourself that way, as a supreme power of sensitivity surrounded by magnificence.

The paradox is this: the greater your sensitivity, the more unbearable the tension. It is much easier to latch onto just one of these allurements, making it the whole. Anyone who grabs a sliver of beauty and insists that it is the whole becomes a fanatic, workaholic, cynic, fundamentalist, or drug addict. To break the tension of living in a universe rich in allurements is to move toward the needless destruction of pursuing a partial vision. The glory of the human is also the difficulty of the human. Precisely because we are able to feel such beauty, we are simultaneously vulnerable to the addiction of fanaticism in any of a million forms.

YOUTH: So you're saying that much of our suffering is because of the powers of the human, because of our possibilities.

THOMAS: That's right. That's exactly the situation. Even the evil actions of human beings reveal the vast and deep sentience that entered the universe with *Homo sapiens*. Humans are especially created to respond to the depths of the magnificent reality of the universe. Therein lies the supreme challenge to live as a mature human.

YOUTH: Then every destructive act comes from responding to beauty?

THOMAS: Ultimately, yes. But an act of destruction resulting from a craving that disregards the whole story and the vitality of the whole is the first link in the chain. Destructive acts are linked through generations as one violence is transmitted and compounded into other violences? These chains of misery can stretch through millions of years, binding up whole societies in torment. In this way, needless destruction is a response to evil that has been handed down. Parents inflict their self-contempt upon their children in physical and psychic abuse, who in turn project their self-hatred onto others and their own children. The Earth suffers under the weight of accumulated misery and pathology, all of which has its ultimate source in acts of egocentric craving. Think of all this suffering, not only human feeling but the torment in so many many realms of the planet! The magnitude of the Earth's adventure staggers the human imagination!

YOUTH: Is there no end to it?

THOMAS: Each individual person has the power of participating in the transformation of the whole Earth. The evil that reaches you after so many millions of years of existence can be absorbed and transformed. You have the power to accept the suffering, to refuse to pass it on to another, to forgive, to end the needless torment, and, most of all, to transmute evil into energy for the vitality of the whole.

The task of maturing into a human being requires tremendous power. It is a matter of authenticity. What powers enable you to achieve your own authenticity as a member of this vast adventure? What powers enable an oak tree to achieve its own authentic function in the living world? What powers enable a star to integrate its processes and initiate its vital creativity?

II: EPIPHANIES OF THE EARTH

Sea

THOMAS: When we reflect on the creativity and forgiveness, the wisdom, insight, and perdurance required of humans in our moment of crisis, we understand the need for the tremendous power of the universe for our work, our survival, and our celebration of life. To become fully mature as human persons, we must bring to life within ourselves the dynamics that fashioned the cosmos. We must become these cosmic dynamics and primordial powers in new human form. That is our task: to create the human form of the central powers of the cosmos.

YOUTH: Wait! The *human* form of the powers of the cosmos?

THOMAS: The same dynamics that created the galaxies created the stars and the oceans. The powers that build the universe are ultimately mysterious, issuing forth from and operating out of mystery. They are the most awesome and numinous reality in the universe. Humans *are* these dynamics, brought into self-awareness, becoming now fully aware of our creative work. We already have these powers in the forms of stars, mountains, atoms, and elephants, but we do not yet have them in human form. We are probing still, exploring, experimenting. Having only just arrived on this planet, we are still learning what it means to become fully human.

We have already discussed the most primordial of these powers, that of alluring activity. There are five other powers central to the creative activity of the universe that are now needed in our task of world-building. These—the powers of Sea, Land, Life Forms, Fire and Wind—are the cosmic dynamics that, when woven together in new form, will show the universe the human person.

We can begin by considering the sea. When I say the sea, I mean one activity of the sea above all: its power to absorb. Water absorbs minerals and draws them into the life of plants, absorbs the soils of the plains, and deposits silt in river mouths. Put a lump of salt in water and it slowly disappears. New York City at the bottom of the sea would also slowly vanish. The sea demonstrates the power of the universe, extant at all levels, to *dissolve the universe.*

YOUTH: What would be another example?

THOMAS: We could consider the elementary particles. When electrons and protons interact with each other, protons are fundamentally and intrinsically changed. We say that the state vector is new, which means that we have a different reality than before. Why? The proton picks up something from its interaction with the electron. This is called quantum stickiness, and is central to the entire theory of quantum mechanics. If the proton is "sticky," it can't just slide by the electron. It absorbs something, assimilating it into its own state of being. It becomes new because, through its interaction with the electron, it has dissolved something into itself.

YOUTH: But it's still the same proton, isn't it?

THOMAS: The situation is similar to water rushing down a mountainside. The water picks up minerals and salts in its journey, becoming something new. When I say it is new, I mean that it enters into new relationships with the Earth. That is how we study the reality of a thing, through its interactions and relationships. If these relationships are new, we have a new entity. An electron passing through hot plasma enters into different relationships; an atom in a highly charged electric field enters into new relationships; so does water passing down a mountain.

YOUTH: But if you wanted to, you could separate the water from the minerals again, right? Then you'd have minerals in one jar, and water in another.

THOMAS: That's true. We tend to define something in terms of the parts it can be broken down into. But that's only half the story. Mineral water can be broken down into its parts, and we can learn something from that. But mineral water as an entity shows itself in ways that its parts can not. Breaking down water itself into its parts of hydrogen and oxygen gives us some knowledge of water, but water as an integral entity reveals things about itself that its parts do not. Learning by analysis has been emphasized over the last two centuries, but we also learn by examining things as wholes.

Notice how this conversation has proceeded: By looking at the sea, we begin to appreciate the manner in which the universe dissolves itself. But when we learn that this activity exists in a different form in the realm of elementary particles, we are assured that we are speaking of reality. This reveals our cultural bias for analysis—the dynamic is as real in the life of the seas as it is in the realm of elementary particles. Each realm has its own integrity; the ocean can not be reduced to elementary particles. If you decompose the ocean into elementary particles, the ocean disappears.

If any event, as we glance at the sea, as we explore the world of elementary particles, we see how the universe assimilates qualities spontaneously. What name should we give this cosmic dynamic? We could call it quantum stickiness, if we wished to keep our attention tuned to the quantum realm. Or we could call it solvency properties of water, if we wished to take the sea, and liquids in general as our reference point. But in order to indicate the universal aspect of this dynamic, we will use the word *sensitivity*.

YOUTH: So protons are sensitive?

THOMAS: They show a minimal sensitivity for each other, yes. The universe is sensitive—it's a realm of sensitivity. Matter is sensitive. To say that an electron is sensitive means that an electron notices things. The electron responds to situations and is intrinsically altered by them. I don't mean that the electron is self-reflexively aware as a human is, however. Perhaps we could use the phrase *quantum sensitivity* to make the same point. All I'm saying is that the electron absorbs something from the world, assimilating it into itself.

YOUTH: I'm confused. This sensitivity, this power of absorbing. . .What are we getting at here?

THOMAS: We're investigating the way humans will mature into their destiny as the human form of cosmic dynamics.

YOUTH: And we've already discussed allurement, and how our destiny is to become allurement. OK. Now it's cosmic sensitivity. But if the *universe* is sensitive, then *we're* already sensitive, right?

THOMAS: Yes. But remember, cosmic unfolding has not ended. If you think of the Earth as forty-six years old, it has only developed flowers in the last year and a half. There is much more to come, but right now the Earth is having difficulty with its most recent creation, *Homo sapiens*. Evolutionary dynamics are blocked until they can bloom in the human form. We are to become allurement, we are to live as cosmic sensitivity, but we're not there yet.

YOUTH: How do humans hold these dynamics back?

THOMAS: Let's consider sensitivity. How do humans fail to evoke this power of sensitivity, this power of absorbing the universe? Let me ask you something: When you see the moon, are you seeing an image of the moon or are you absorbing the moon? That is, what happens when you glance up at night and see the moon?

YOUTH: Well, the light from the moon comes to me and hits my retina and I get this awareness of the moon.

THOMAS: So seeing the moon is like watching a television screen that has an image of the moon, right? It's there for a while, then it's gone.

YOUTH: Well, yeah.

THOMAS: Now, in actuality, something much different happens. When you look at the moon, you are absorbing the moon just as the ocean absorbs minerals.

In terms of quantum mechanics, you as an

individual body are represented by a particular quantum state. This includes the interactions of all the elementary particles of your body. Now imagine a patterned wave of light flowing into you. Some of the photons of this light wave interact with your own elementary particles, and through this interaction your quantum state is changed. This is the quantum "stickiness" we discussed before. Your particles are new in the sense that they have absorbed something from the photons and entered a new state of being.

Imagine a great number of tiny bells hanging near each other. If some of these are struck sharply, they will transmit their own resonance throughout the ensemble. No bell will remain the same, thus creating a new state for the whole of them. The same situation holds true for your body: interaction with the photonic shower creates a new quantum state.

This means that when you stand in the presence of the moon, you become a new creation. The photon's interactions have entered into the quantum state of your entire ensemble, and you are, through these interactions, a moon-person. It is not something you *have,* an image or an object, so much as it is something that you *become*. The elementary particles of your body have absorbed an influence and in that sense they—and you—are brand, spanking new, a human being resonating everywhere with moonlight.

There is no separate self "having" this image; rather, your totality is permeated with the moon's presence, and this totality, in reflecting upon and within itself, exists in a new awareness: the awareness of the moon. You are the self, you are the moon. Then there is only self-moon. That is your reality. That is what cosmic sensitivity means for the human.

To develop the power of cosmic sensitivity is to understand that to be in reality means dissolving the universe, absorbing it into your new self. To be is to dissolve and draw up, to *be* dissolved and drawn up. The universe is a hard yellow candy, to be sucked on and swallowed until it dissolves, and, in that moment of dissolution, we emerge. A hardened mind can not respond to the presence of the moon. The moon's riches can not be tasted, so the moon can not show itself. The interaction between the rigid person and the universe is superficial, because the sensitivity is dim.

YOUTH: So we cripple our sensitivity by thinking that we are separate selves that "have" these images of the moon or whatever?

THOMAS: And by assuming that our feelings are just our feelings! Do you see the mistake here? The human awareness could never know the throbbing presence of the moon and all the intensity of feelings *were it not for the moon itself.* These feelings are as much the creation of the moon as they are of the human. We partake in the great presence of the night sky, awareness rising out of the interaction. Our sentience, our feelings of wonder and awe, emerge out of the universe. We could not feel awe without the grandeur of the universe. These profound feelings are not just ours; they are the universe reflecting upon itself.

The moon and you conspire in this moment of intensity. Remove yourself or the moon and the reality evaporates. To live is to enter this beauty, surrounded by enchantment, summoned by magnificence. When we discover awe, we enter an enchantment that enjoys a supreme objectivity. The universe is enchantment.

I have chosen the moon, but you have your own moments of beauty. In each case you can see that the beauty and feelings of the universe welled into you in that moment. Each reveals the cosmic sensitivity that exists within the human form. Protons respond to particulars of the universe, the seas to different qualities. The human's sensitivity enables the beauty of the universe to be caught in self-reflexive awareness. The glory sweeping through the universe is glimpsed in each instant of your awareness of beauty.

Another image might help you here. A hundred television programs surround us right now, yet we see nothing. There are men and motorcycles and whales, young women and sail boats, all right here, flooding us, but we remain unaware of them until we create a unit that can evoke them from electromagnetic waves. So, too with the deep feelings of the universe. They sweep through the cosmos, unnoticed by humans who have not developed their innate sensitivity.

YOUTH: But what do I do to develop sensitivity?

THOMAS: Learn to listen. You have to devote yourself to this skill over long periods of time. *Really* listen. The magnificent feelings of the universe flood you: listen for them in every situation of your life. Listen to your friends with such sensitivity that you are leeching the very air that surrounds you. Listen so that, if it were given to you, you would hear the whirring of Saturn's rings, or the least wind a continent away. When you leave your friend, her presence will emanate from you. Notice this, feel this presence radiating away from you so that you can deepen your awareness of the way in which you dissolve the universe and absorb it.

When you walk into a forest, learn to tremble with the magnitude of what you are about, and you will never walk out. There will no longer be that self that approached the forest, for you will be new, you will bear the presence of the forest with you. Forests are alive with music on all sorts of hidden levels, and when you hear this music you will know that the forest has permeated every cell of your body. Sip a cup of coffee the next morning, and all the fir trees grow warm. The natural, human, and divine worlds flow together into our feelings. You need no teacher. The universe is your teacher, the forests are your teachers. You will know when you fail to learn, for failure is punished with boredom. If you develop even the least flicker of sensitivity, the universe will come alive within you.

Think of how, for billions of years, the presence of the fireball flooded the Earth. There it was, each instant, washing over everything on the planet, and only now have we learned to become sensitive to it. We are awash with the presence of the universe, already swamped in its beauty. All things have discharged themselves into the world, merely awaiting our development of the sensitivity to respond to them. To live as a mature human being is to journey home, and our home is enchantment.

Land

THOMAS: Now we can consider land: the soil, rocks, mountains, continents, and elements—matter. In particular, I want to tell you how land shows the power of linking back, of re-connecting, of re-membering. The cosmos remembers in its own way, and we can see this most clearly in the land.

The elements themselves are frozen memory. They present to us the work of supernovas billions of years ago. A sturdy memory, certainly, when we have difficulty keeping track of our phone numbers. But the shape of the elements has been preserved and remembered through all these eons. The elements show us the original form given them at their emergence into the universe.

The crust of the Earth holds the story book of life's adventure, especially of the last six hundred million years. The gneissic rocks of Greenland capture in their crystalline formation the story of our Earth four billion years ago, when Earth was just leaving its molten state. The journeys of the continents as they crashed against each other and floated across the oceans on the spongy rock of the mantle has been recorded in the mountain ranges, seas and trenches left behind by the collisions.

The cosmos desires to remember, but it does not always succeed. On Earth, this cosmic dynamic of memory has succeeded with such a diversity of re-membering that even two-legged creatures plagued with tonsilitis and tax worries have evoked the great story from the rocks.

YOUTH: How have we failed to develop the cosmic dynamic of memory then?

THOMAS: Our understanding of memory is anthropocentric, to begin with. We have limited ourselves unnecessarily.

YOUTH: What do you mean?

THOMAS: Your arms are memory poured into flesh, fiber, and bone. Do you understand that?

YOUTH: No.

THOMAS: Consider a mountain goat. These animals have the ability to stand on a tiny ledge of rock with the wind blowing and the rains crashing down on them. Their hooves, in particular the outer shell of the hoof surrounding the inner pad, allow them to get a hold on a small rock almost as if they were grabbing it with pliers.

What we have to appreciate is that this adaptation required millions of years. The ancestors of present-day mountain goats lived on mountains, adjusting to the mountain's shapes, the difficulties of gravitational pull, and everything else. Those shapes that were most successful in fitting into the mountain's reality were selected for survival, so that what we see now contains all that previous experimentation. The hoof is the memory of the ancestral tree. It didn't show up accidentally; it was shaped by the accumulated experience of millions of goats.

The point is, matter remembers the elegant hoof. The genetic sequence enabling such a hoof to be fashioned becomes dominant in the gene pool, passing the hoof around to all members of the species. So, you see what I mean when I say the hoof is permeated with memories from the past. From this standpoint, the hoof *is* those memories.

YOUTH: But how does this pertain to the human?

THOMAS: Just as hoof is memory, the human body is memory. Think of how many creatures are involved in the ancestral tree required for the creation of our fingers! When you lift your hand, you are lifting all the vast experimentation that led to that hand. There before you is the history of the great events of the universe: the biological exploration, the supernova explosion, all the significant moments of the last twenty billion years are remembered.

YOUTH: But what does the remembering?

THOMAS: Matter. Matter in the form of molecules. The sequence of molecules that make up your DNA is a sequence of memories. Do you see how the cosmic dynamic of memory relies on a particular event to show itself? We can't *see* memory's dynamic any more than we can *hear* allurement's dynamic; we can only look with astonishment at the genetic sequence of molecules captured—remembered—by the DNA in all cells.

YOUTH: Why is there memory?

THOMAS: We are investigating those powers of the universe required for its creativity, for building its astounding events. The universe remembers so that it can benefit from the labor and awareness of previously existing beings. Why should it forget moments of tremendous cosmic or geological or biological beauty? Think of how many billions of creatures were involved in the accomplishment of the animal eye. What a tragedy if this were not cherished!

YOUTH: How do we develop the power of memory in this cosmic sense, then?

THOMAS: Begin by thinking of memory as an activity. Re-membering is something that the universe *does*. For the cosmos, memory is the way the past *works* in the present. The universe doesn't want to waste anything. If it can get the past to do the work in the present, well, why not?

Consider the oak tree. Proto-oak trees emerged some two hundred and fifty million years ago. All that work, all that probing creativity, patience, and suffering that went into the creation of the oak was present in the acorn that unfolded into this great red oak here. History is captured by the acorn, so that when placed in soil, watered, and surrounded with air, the oak unfolds all the beauty that was enfolded into the tiny acorn. Think of the minerals moved this way, that way, by and through the oak tree as it emerges. What guides all that pushing and activity? What guides the decision to branch here and not there? What decides to avoid the dead end and choose instead this already tested and reliable course of events? The oak tree as a whole does—yes—because the remembered past is there, guiding, influencing, choosing, and affecting the development of the whole. In this sense the past is present, actively working in the growing oak.

Modern humans don't understand this. We regard history as something dead and gone. We live in the fingernail of the present, and fail to realize how this cripples us. At best, we think history is six thousand years old, something involving humans only, most of it irrelevant. We have convinced ourselves that our ancestors all tried to be like us, but failed. We think they too would—if they could—devote themselves to a machine world, a bigger and bigger GNP, and a continent filled with consumers.

Can you imagine what would happen if the plant world imitated us? If they regarded the creations and the lifestyles of their own ancestors as outdated, beneath them, not worth remembering? They would snub the lower forms and their photosynthesis, trying to do without that stupendous and permanent achievement of harnessing the sun's light. If the plants imitated us, life on this planet would wither in a week. So the question we should ask ourselves is this: What would happen, on the other hand, if we tried to imitate the plant world? What would happen if we began to see that the achievements of our ancestors are permanent creative advances, handed down to us for our benefit?

To begin with, we would cherish the tribal peoples of the planet, just as plants cherish photosynthesis. These peoples have lived with the rhythms of the planet for tens of thousands of years, fashioning skills and procedures over the ages that must not be lost in our rush to pave the Earth asphalt. They established basic attitudes toward the primary realities of this world, and fashioned traditions that have endured difficulties that we can hardly imagine. They forged rituals and initiation rites to remind them that this Earth is sole provider and sustainer of all life forms. To develop our cosmic dynamic of memory means that we assimilate the wisdom of all native peoples. They have accumulated knowledge that we can not do without, wisdom that, should we lose it, we could never reproduce.

The achievements of the great classical civilizations must also be remembered, for they are permanent acquisitions of the planet, as crucial and as irreplaceable as tribal processes. During the times of the classical civilizations, humans remembered being. They were able to face the tremendous awe still found in all of Earth's religious and poetic utterances. It was then that humans first grasped, in a well-developed conscious awareness, the deepest reality of cosmic memory, for they realized that the universe cherished the creations of time, never losing anything of beauty. Because we have forgotten these insights, we industrialists suffer crippling fears of death. Instead of sinking into the joy of living, we bury our lives in trivia and pressurized distractions, anything to forget the fact that we are alive, in being, and destined for infinite delight.

YOUTH: Is this what you mean when you say we block cosmic dynamics when we forget, when we think we have no need of the past?

THOMAS: To forget the past! What can this really mean, other than to deprive ourselves of infinite power? The universe desires to break through into human form, but we cripple ourselves, insisting upon living in a fingernail's edge of our true heritage. We are just like some ignorant oak tree who insists upon ignoring all the efforts of the past, setting out to make its own leaves and own form. Impossible.

YOUTH: But when are we like this?

THOMAS: In nearly everything. Even the most ordinary tasks show this. Take eating: Our relationship toward food is simply wrong. Instead of eating the natural foods Earth has created over eons of subtle experimentation, we stuff ourselves with fake junk put out by multinationals with less knowledge of the Earth than could be stuffed into an empty peanut shell, resulting in cancer, heart disease, and all the needless suffering associated with folly. We need to realize that, from a biological point of view, eating is remembering. Why? Because food is rich in the information our bodies need. Through hundreds of millions of years, life forms learned to feed on each other. This means more than supplying fuel. It means supplying the informed sequences of molecules and amino acids required for our epigenetic unfolding. Our bodies *wait* for, *expect* a particular spectrum of foods. Not just anything will do. Particular molecular compounds are required, those that were fashioned by the millions of years of creative experimentation.

YOUTH: But how is eating remembering?

THOMAS: Many of our physiological patterns of activity depend on certain complex chemicals provided by natural foods. The physiological processes are the way the body remembers its ancestral heritage, and this heritage insists on particular natural foods for its remembering. When you eat grains, legumes, and good, fresh meat and vegetables, you enable your body to remember its powers.

It is similar to what happens when you leaf through an old photo album. The pictures key all sorts of memories, and you are flooded with the past coming alive within you. That's what eating is like. The foods enable patterns of activity to start

up. If we understood that food was memory,
would stop our miserable eating habits.

YOUTH: So remembering includes eating habits. Eating is a form of memory. What other ordinary activities are memory?

THOMAS: Exercise. To exercise actually means to bring into action. When we exercise, we bring into action our ancestral memories. Our bodies remember that we lived in trees and forests. We need to crawl and climb and run if we are to develop our intellectual, emotional and spiritual capacities. We did not emerge from an austere iceberg of a distant planet, but in the particularities of this Earth and its forests. When we wander through the mountains, climbing and running, our bodies remember those deep patterns of behavior intrinsically tied to all that we are. We tend to think of exercise as losing weight, as trimming off the fat. But to exercise is to enable the body to remember its past, so that it can stretch out with all its intertwined powers of being and thought and reflection.

To remember is to know. To remember the great events of human history is to know them. We remember the tremendous creativity of Earth's history when we know the subtleties and complexities, the overall coherence of these stupendous events. To develop the power of memory is to deepen our knowledge. The difficulty with the word knowledge is its connotation of *self-reflexive* knowledge. We have belittled the word in our dualistic, anthropocentric usage. Knowledge is memory; to know is to remember. An engineer knows how to build a bridge because he or she can make present those procedures that have worked in the past. Thus we see that our knowing and our remembering are present in the animal and plant worlds as well.

You can understand, now, what I mean when I say we need to remember the universe. We need to study the cosmic story, the Earth story, the human story, until we know it in its essential forms. A person who does not know the story of the universe is not yet living up to human destiny. But this knowing is not only cerebral; to know the story of life includes eating natural foods; to know the story of human civilizations means feeling the profound intuitions they achieved; and to know the story of the universe means to allow the great, numinous past to come alive in your present being.

YOUTH: You know, this is so different from everything I was taught. I have never once thought of studying history in this way, so that the universe could come alive in me.

THOMAS: I realize that. The switch out of an attitude where the human is the center of everything, to a biocentric and cosmocentric orientation where the universe and the Earth are the fundamental referents, is *the* radical transformation that we are presently involved with. It *is* disruptive. We are so quickly confused because we are accustomed to forgetting the Earth and cosmos to concentrate on the human world. But when you begin to grow into this larger way of living, you will discover new freedom, and a vast vision of being that makes the struggle worthwhile.

YOUTH: But it's so confusing, especially when I think of where to start.

THOMAS: Remember beauty and awe. Remember the stupendous achievements of our unfolding universe. Begin by memorizing that sentence, if you like. And begin at home. Remember the moments of beauty that you have already experienced. Reflect on your own life. What were the central moments to be remembered? Carve them on the inside of your belt, fashion tapestries with the major events celebrated, or paint symbols on your wall that will call them to mind. Put down a paragraph of language for each of your central moments of living on this planet. These are to assist you in your present. They are to empower you! Bring all these moments of awe, or difficulty, or endurance, or nobility into your present, and you will be already remembering in the way that you must.

Leonardo da Vinci understood something about the work of memory. If he was captivated by a face, he would follow the person around for an entire day if necessary, watching and studying and drawing. Not until he could reconstruct the face without looking would he be satisfied. This is what is meant by the beautiful phrase, *learning by heart*. He knew the face, and he had etched it into his being, into the deepest reaches of himself. He remembered the face because he had identified with this beauty so deeply that it lived through him. When we remember beauty, we become this beauty for the living present.

YOUTH: But what about evil and sadness? Are these to be remembered too?

THOMAS: Yes, in a different way. In general, we remember so that we can ignite and enhance life.

This requires of us that we remember suffering, pain, and hardship. In fact, the body remembers damaging events of the past even though these are often not remembered in self-reflexion. The body remembers so that the mistake can be avoided in the future.

It is essential, then, to remember evil. Feelings of guilt stem from this. The universe insists that we pay attention and remember our pasts, whether as individuals, societies, or species. We can be plagued by memories of error until we turn them over in our minds and penetrate them with clarity. Once we have snatched understanding from their core we are released from our guilt; the lesson has been learned, the impasse overcome, and the spontaneity of creative activity returned, refreshed and effective once more.

We can not approach any satisfied existence as humans until we engage our capacities in this great work of remembering. We are uniquely fashioned to succeed—the dynamics that forged the fireball and the trillions of stars are at work within us as well, bringing into a single flaming intensity the whole vast epic of reality. What we eventually discover in our passionate remembering of the galactic, terrestrial, biological, and human stories is that a study of the universe is a study of self.

LIFE

THOMAS: The sea embodies cosmic sensitivity for us; the land, cosmic memory. Now let's look at life forms: What's the first thing you think of when you think of life?

YOUTH: That's easy: death.

THOMAS: That's what comes to mind?

YOUTH: Well, everything born is going to die.

THOMAS: That's not exactly true.

YOUTH: What?

THOMAS: Some creatures don't die. In fact, for two billion years the life forms on the planet Earth did not have death as an inevitable ending to life.

YOUTH: I don't understand.

THOMAS: For billions of years, death was not a biological necessity. Nothing died "naturally." The earliest creatures might be killed, crushed, or starved to death, but death as we know it was not inevitable.

YOUTH: They could live forever?

THOMAS: If the conditions were right, yes. The bacteria present on Earth today are like that. In fact, the prokaryotes alive on Earth *now* could possibly have existed at the very emergence of life— some of them might be four billion years old. We don't know, of course, but it is a possibility. The point is this: Death was an invention of evolutionary creativity. Life does not inevitably mean death. In the beginning, death simply wasn't necessary.

YOUTH: What a lousy deal.

THOMAS: You're disappointed?

YOUTH: Can't we get back to that earlier way of living?

THOMAS: Would you want to?

YOUTH: Well, of course.

THOMAS: You think the universe was mistaken when it invented death as a biological inevitability?

YOUTH: What's gained by it?

THOMAS: Good question. Why would life create biological death? Let's start at our end: suppose we eliminated natural death. The first thing that would happen would be the need to eliminate reproduction. Existing humans would, of course, want to remain among the living, and once the continents were jammed, we couldn't allow for any newcomers.

YOUTH: All right, that sounds fine.

THOMAS: It sounds all right for a while. But what about a million years later? The same old humans still dragging around the planet, the same old animals—pretty dull. The sad thing is that the anxiety over death would be a billion times worse than before, for we could still die in accidents. You'd never risk a single step outside your little hut! Why risk an eternity of life on Earth for anything the least bit dangerous?

Even the animals would figure it out eventually; they too would huddle in caves. Who knows how far this terror would spread? Perhaps one day the sun would figure it out as well, stop burning, turn dark, and huddle inside itself forever.

Life avoided that dead end of paralyzing

immobility, devising instead its greatest desire and accomplishment: novelty and surprise. Look at the adventure of life all around us! Adventure, surprise, risk, and excitement; these are the fundamental desires of life.

YOUTH: Why couldn't we have that without the anguish of knowing we're going to die?

THOMAS: You're being self-centered. It's not painful to live as a squirrel or an elephant, even though they too are going to die. They don't drag around all day with long faces. Nor do they spend lifetimes writing dreary novels about existential angst over death, or, worse, inflict a miserable fear of extinction upon everybody else. It's not difficult to live as a majestic sequoia or pungent prairie grass or the delicately industrious hummingbird.

YOUTH: But they don't *know* they're going to die! Why couldn't life stick with forms that were ignorant of their death?

THOMAS: An excellent question. Why did life create forms that would become aware of their own individual deaths? Let's approach this question from the point of view of the emergent cosmos. The question then becomes, "What is gained for the unfolding cosmos to include particular creatures —humans—who are aware of their own deaths?"

Why make us aware of our deaths? To deepen the adventure of life; to underscore the drama of each instant. The universe desires to show itself! The universe is a showing of the unnameable mystery out of which being shines forth. How else could the universe feel *its* own staggering value? How else, but through a human space aware of its own individual end? Within human self-reflexion can be felt a glimmer of the supreme preciousness of being, and we would certainly not be able to feel this were it not for our awareness of death.

YOUTH: So that's why we have to suffer the anguish of knowing our death.

THOMAS: It's difficult, yes. It's the task set for the human, and we suffer tremendously in our role of carrying the awareness of life's precious and fragile beauty. But our reverence is our gift to the universe. Who can feel the stupendous, fragile beauty of the great black whale soaring through the ocean but the human? We are able to feel and cherish the infinite significance of the whale as it dives into the freezing ocean, to value it for what it is. That is our gift to the living world, to *see* it, to feel the moment, to speak it, to celebrate its truth. You might say the whale has it easy. The whale is free from the anguish of self-reflexive awareness of its approaching death. But because of that, the whale can not feel its own beauty, its ephemeral magnificence; that is what the human must do, or our suffering is in vain.

YOUTH: But we die anyway, and our feelings end. Our feeling of the whale's beauty and preciousness disappears.

THOMAS: It disappears in time, yes.

YOUTH: That's what I'm saying.

THOMAS: You're locked into identifying with the temporal. Time is not the fullness of being. There are existents and there is emptiness. Both are real. The eternal, the transphenomenal, shows itself in time, yes, just as the dynamics of the cosmos show themselves in concrete events. But what is invisible is real as well.

YOUTH: Then where is the feeling of the whale's beauty?

THOMAS: Captured, cherished, and remembered. The universe never loses anything of value. Relax! You couldn't hang onto that beauty anyway, no matter how hard you tried. Do your work, and the cosmos will do its work.

YOUTH: I hate thinking I'll just disappear at death.

THOMAS: If you surprise the world with your life, the world will surprise you at death. Don't think of death as extinction; such uninspired speculations are simply much too prosaic to be true. Your dull imagination insults the very grandeur and staggering wonder of this universe. It's alright to be immature, but don't project your callow views upon the universe. Yesterday you knew nothing about the primeval fireball or the stunning dynamics of the star's unfolding, and yet you feel qualified to say that the universe was wrong in creating death?

Rather than hiding from your death, or repressing your fear about death, embrace your death. It will serve you.

YOUTH: How?

THOMAS: By enabling you to show yourself. Precisely because you are aware of the limits of life, you are compelled to bring forth what is within you; this is the only time you have to show yourself. You can't hold back or hide in a cave; you can't waste away in a meaningless job, cramming your life with trivia; the drama of the cosmic story won't allow it. The supreme insistence of life is that you enter the adventure of creating yourself. Each instant of your life has folded into it unnameable significance; all rests on your self-creativity, for out of you comes forth ultimate reality. The dynamics that fashioned the stars are now brought into your self-reflexive awareness, and what they create is your free adventure, your surprise for the universe.

Yes: death is terrifying. Do not belittle it. Do not try to reduce this. Do not project your puny ideas upon it. But *use* death's awareness as you would a fuel or a lamp: as a secret guide who will lead you into the unknown and mysterious caverns of your self so that you can bring forth what you truly are. Your creativity needs your awareness of death for its energy, just as your muscles need long and painful workouts. Cherish your awareness of death as a gift to you from the universe. If you did not have this way of seeing the infinite significance of each moment, would *anything* have the power to get you out there to live your life?

What is especially exciting about our own time is the vision of the death of the species, and of the planet as a whole. Frightening, terrible, horrible—yes, certainly. But this is exactly what has the power to ignite the deepest riches within us. We can no longer live within the previous world-picture. We know that we have to do something, create and change in the essential dimension of things. The terrifying vision of an Earth gone black is psychic food for the human species. It brings us the energy that we need to re-invent ourselves as the mind and heart of the planet. We now take our first steps into the planetary and cosmic dimensions of being, moving out of the anthropocentric modern period and into the cosmocentric, unfolding universe.

YOUTH: But what does it mean to become the

mind and heart of the planet?

THOMAS: To live in an awareness that the powers that created the Earth reflect on themselves through us. That's why we are discussing the night sky, the sea, and the land. Each of these reveals cosmic powers that we are to have and become. We are to live as alluring and remembering activity, as shimmering sensitivity. And this means the cosmic dynamic revealed by the life forms: surprise and adventure. Call it play; adventurous and surprising play. That's what life reveals; that's what life *is*.

YOUTH: And that is what we are to become?

THOMAS: Yes. But, again, we must understand something especially important. The insistence that we become adventurous play is not our insistence alone—the universe insists on it. As in each of the previous cases, the universe created our sense of adventurous play as the latest extravagance in a long history of advancing play. By enhancing it, we work with the grain of cosmic dynamics. Do you see what I'm saying?

YOUTH: We're enhancing the movement of the universe, then.

THOMAS: Yes. Life showed surprise from the beginning. The earliest organisms advanced by a random appearance of novelty. We call this genetic mutation totally random, by which we mean that there is no controlling machine. The genes show a fundamental freedom of activity. Nothing could predict the outcome before the appearance of the new form of life.

This presence of free activity was enhanced with sexual recombination. Now entire complexes of possibilities could be played with, rather than only individual units. The adventurous play of the life forms bursts into the bewildering and sublime diversity of the past five hundred million years. All of this profusion of being and beauty is the outcome of play, of risk, of surprise. The creation of new life forms is not determined, but is the outcome of life's intrinsic freedom.

Nor is play's manifestation limited to activity at the genetic level. Life forms play, especially the young of species. In mammals there is a recognizable difference between the young and the old, not only anatomically, but behaviorally as well, and the most conspicuous behavioral difference is the propensity and ability to play. The young come into the Earth's system of life as if play were what they were created for. They explore, cross the normal limits of things, leap about without reason, climb too far out on limbs, and fall in the water when their curiosity fastens on something new and strange there. The young reveal the core of life's mystery: the need and opportunity for adventurous play.

Now let's consider the human in all this. Biologists have discovered that in the primate order there is little genetic difference between species. The chimpanzee and the human share over 98 percent similarity in their gene pools, an astonishing discovery when we consider the tremendous differences between these species. But what is the *essential* difference? What line was crossed that created the human form, and not before? Is it the size of the brain? Current thinking locates the difference between humans and other primates in the ability of the human to make play its dominant activity throughout a lifetime. Unique among species, the human makes exploration, surprising discoveries, experimentation, and—above all—learning the central activities of life itself.

The human form of life can be considered the child of the Earth. This is especially clear when we examine the anatomies of other primates. The head of an infant chimpanzee resembles the head of an infant human in size and shape, but as the chimpanzee reaches adulthood, its head changes in significant ways. The human head remains comparatively the same infant head, only larger. In fact, the infant chimpanzee's head looks more like an adult human's head than its own future adult shape. This dynamic, in which the qualities of the young are retained into mature stages, is called neoteny. We can then begin to understand the human as an eternal child. The first human types were young primates who never "left" their youth. The shapes of their juvenile bodies were retained into adulthood, as was their youthful behavior. The great accomplishment of the human form, then, was the creation of a mature form of childhood, a form of life that, upon reaching adulthood, could continue to devote itself to a lifetime of adventurous play.

So you see what I mean when I say that life insists we develop the cosmic dynamic of adventurous play.

YOUTH: And if we don't, it means that we are blocking life's unfolding again, right?

THOMAS: Yes, that's our impasse now as a species. We can think of it this way: Each species has its own habitat, that place where the species can flower forth. If a species cannot find its proper habitat, its true powers of life cannot be evoked. A species denied its habitat perishes; we see it all around us. What is the true habitat of the human? *Adventurous play.* A human denied this habitat of adventure and surprise and play is denied the opportunity to become truly human.

Our anguish today rests in our failure to recognize our true talent. We thought we were supposed to become full time consumers in one great world-wide consumer society. But that brings no satisfaction, and we end up trashing the garden spots of the planet. We tried to live as appendages to our machines, discovering only unrelenting meaninglessness in the midst of grime and noise. What else could we have expected, trying to live outside our habitat? Can a whale live in hydrochloric acid? Can an oak tree send down roots in a tar pit? We will finally move into our destiny when we understand that we are to live in and as adventurous play.

YOUTH: What would this mean, specifically?

THOMAS: Who knows? That's the great thing! We can't go to any other species and ask them. That's the whole point! Adventure is an adventure into the unknown. True play is without predetermined direction or definition. We are to explore, to learn as deeply as we can, to probe and experiment, and above all to laugh. Humor already reveals the presence of adventurous play—a deep belly laugh might be the one true cry of the human being.

Don't be saddened by our lack of knowledge concerning our destiny as cosmic play become self-reflexively aware. Put your trust in the entire cosmic process. It has labored through twenty billion years; believe me, you are well-equipped for the job. Think of the tremendous labor of all living forms to have finally arrived at you, the ultimate child of the planet. They did their work; now you do yours! Plunge into the work of living as surprise become aware of itself. You are the essence of surprise, the heart and core of play. Show yourself as truly as you can, and you will in that moment shine with the freedom and frolic and fecundity of creative play.

To say that play is essential to the human species is to corroborate what creative scientists, artists, and the great saints have understood as central to their own activities. Play, fantasy, the imagination, and free exploration of possibilities: these are the central powers of the human person. The development of the Earth depends on the development of the human into its destiny as the self-portrait of adventurous play. Who can say? Perhaps all the other species are capable of profound playful exploration of relationships as well, only awaiting us to start the process. Perhaps the entire natural world is a tremendous party, a festival, and we the long awaited champagne.

Fire

YOUTH: You know, I was just listening to you and I realized something very strange: I was excited, and I thought of becoming a Master of Play. I don't even know what that is, but I thought, wouldn't it be wonderful if there were schools somewhere that taught humans how to become true masters at the art of play?

Now what was so strange was that this idea didn't appear strange. Do you see? Never in my life have I considered such a thing, and if anyone suggested it to me I'd think he were crazy, but here I was actually thinking about it in a serious way. Isn't *that* strange?

THOMAS: That you would come up with the idea of becoming a Master of Play?

YOUTH: No, no, no. I mean, which was the real me? The one before who would have thought the idea was just crazy, or the one now who thinks it's a possibility?

THOMAS: Which you is your true self?

YOUTH: Yeah.

THOMAS: Well, the strange thing is that your true self is never the content of whatever you happen to be thinking or feeling. No matter how well you get to know your self, you will never be able to say what this self is. We create images of the self all the time, but they're just images.

Think about it a moment. If you could say what this self *is*, then what is doing the saying? Something behind the very speaking or understanding.

YOUTH: This is confusing.

THOMAS: Your self is an organizing activity. Your self is not that which is organized and given form. You just said that you are pondering this possibility of Master of Play. This new synthesis is offered up by the self-organizing activity that you *are*. The self can never be captured completely in language, ideas, images, or creative productions. On the contrary, it is the self that *is* the very doing of all these things, the power of fashioning all these things.

What we need to do is to look at the universe as a whole, examining the self within the context of the emergent universe. We are involved with the discussion of another cosmic dynamic, one best revealed in the presence of fire.

What is fire, anyway? Consider a burning candle. The wispy black smoke rising steadily above the yellow-orange light; the wick, black at the top, white beneath; the wax, liquid on top of the candle, solid beneath, vaporous around the wick area. What is the flame? Is it the light given off in all directions? The wax as it combines with oxygen? The chemical products resulting from this combination?

YOUTH: Can't we just say that the flame is all of those things taken together?

THOMAS: No, not exactly. They show the flame's activity. But consider this. If I alter the thickness of the wick and the composition of the wax, the entire display alters. The color and temperature of the light and vapors will change. Yet we still recognize it as a flame. The flame organizes all these different materials into its own persisting process. A flame is an image of unseen organizing activity.

We say that the flame *is* this organizing activity. It shows itself by organizing materials in a particular way. If the materials are altered, the various conditions changed, the flame can still show itself with these new materials. The flame is an activity, a self-organizing power that spontaneously erupts and shows itself whenever it is able to.

You are similar to the flame. Here you are in the universe, and so long as particular conditions obtain—food, air, beauty, interest, the promise of adventure—you leap into existence. You show yourself in the manner in which you organize the materials of your world: your ideas and feelings, body and relationships. You are not just the things you do, not just the thoughts you think, the convictions you hold. You are that which shapes all of these. You are a power creating the whole complex work of art that is your life, your manifestation in the world.

YOUTH: And now you're saying that this self-organizing activity is another cosmic dynamic?

THOMAS: That's right.

YOUTH: And this is found primarily in the higher animals?

THOMAS: It's found everywhere; that's why we call it a *cosmic* dynamic, and why I chose fire as revealing this dynamic. When we are in the presence of flame, we are in the presence of unseen organizing activity.

YOUTH: Are there other examples of this?

THOMAS: Of course. A tornado is an extremely powerful presence of self-organizing activity. The tornado continues, whether it's over a desert, farmland, or the ocean. It organizes all sorts of materials into a process that we recognize as a tornado. It resists attempts to break it up, just as you would resist attempts to break up your existence. The tornado, the flame, you: each an organized process in which parts are drawn into a whole, and the whole itself is involved in this persisting activity.

YOUTH: You talk as if the tornado has a self.

THOMAS: What is a self?

YOUTH: *I* am. I'm aware of what I'm doing.

THOMAS: You are self-reflexively aware, yes. But a week-old infant is not. Nevertheless, we regard a human infant as a self.

YOUTH: Then what is a self?

THOMAS: Unseen shaping. "Self" may not be the best word for this activity, but we need to bring to mind the autonomous centers of activity found all about us. A flame is unseen shaping, as is a tornado. They exhibit a fundamental dynamic of the universe that we tend to overlook.

That is why I prefer to think of the flame as a self. A tree is a self: it is unseen shaping more than it is leaves or bark, roots or cellulose or fruit. The tree, as a self, organizes all these millions of operations so that it can enter into relationships with air, rainfall, and sunlight. What organizes all these materials? Is it something we can see? Something we can grab unto? No. The same with humans. Who organized your thoughts so that you were suddenly contemplating becoming a Master of Play? We can not point to anything physical and say, "*There* is the self!" This holds for the tree's activity as well as for the human's. What it means is that we must address trees. We must address all things, confronting them in the awareness that we are in the presence of numinous mystery. Who shapes the tree? Who shapes my thoughts? We are in the mystery of the self.

YOUTH: But the tree doesn't know I'm addressing it.

THOMAS: How do you know? But I'm not worried about the tree here. I'm wondering about humans. *We* need to address trees and all things. *We* are the ones who forget awe, mystery, and the astonishment of being. We think of trees as firewood, as pre-plywood, as unvarnished hardwood floors. We've convinced ourselves that they are just inert stuff, standing there for twenty years until we get around to cutting them down. We're deluded.

We must face the mystery of unseen shaping. The tree has its own destiny in the forest, its own life, its own fate; one that takes place entirely outside our little two-legged projects. The tree is made of the materials of the same super-nova as you. It drifted through space right along with you, mingling with the very elements of your body. And now it exists in the form of a tree, with its own hopes for rainfall and sunshine and all that it needs. It knows exactly what it needs, and if these are supplied, it leaps to the task of showing its presence. If it can not get what it desires, it certainly suffers, shrivels up, and dies.

YOUTH: I can't imagine what I'm supposed to say to a tree.

THOMAS: Don't say a thing. Close your mouth in

the presence of the trees. Just think. Stand there and think to yourself: "There you are, tree. Growing, developing, enjoying rain and sunshine and soil. I don't know what your life is like, standing out here all night in the snow, owls scraping your bark; or standing there soaking up tons of sunlight. I don't even know what it's like to hope for sunshine or stand helpless with a forest fire blazing your way. I can't even imagine the joy you might feel when the sun shines and all your vast processes of life churn forward, throwing out tons of water, producing such complex seeds. But whatever destiny you are to live, I want you to go on living it. Whatever relationships you enjoy with the Earth's vitality, I want you to go on enjoying them. I don't know what my own destiny is, nor what relationships I will enjoy in the future, but together we make up part of this vast mystery of the Earth's life, and that is enough for now."

Remember that the tree doesn't need to be addressed. It is *you* that needs to address the tree. You are the universe pressing into an awareness of itself. It is your task to become deeply aware of trees, and of all that is.

YOUTH: OK, a human is a self, a tree, a flame. . .

THOMAS: *Every thing.* There is no thing that does not exist outside unseen shaping activity. Consider an atom. No one has to teach an electron about s- and p-orbitals. Bring electrons, protons, and neutrons together in the right conditions, and you see the helium self appear. Unseen shaping organizes the particles into the stable community of the helium atom or any other atom. If these atoms then experience any number of stressful situations, they adjust so that they can maintain themselves. They absorb or radiate energy, alter and organize themselves so that they perdure. That is what a self does: self-organizes. Atoms and flames, tornados and trees, each presents a centered,unseen,shaping dynamism.

YOUTH: But why didn't I learn that an atom had a self?

THOMAS: We never knew much about atoms. Newton spoke of atoms as dead massy balls. Dalton hypothesized the existence of atoms in the nineteenth century to explain various phenomena. It was not until the twentieth century that we got close to the life of an atom, and for the last decades we've been so enthralled by the dynamism we found there that we never stepped back long enough to recognize the similarity between an atom's self-organizing powers and that of say, a tree.

A similar situation holds in our relationship with the Earth. We have never had the opportunity to study the whole Earth in an empirical manner before. Only now are we finally realizing that the Earth is a self as well. The Earth is a self-organizing process of astounding complexity and achievement. It's a matter of intimacy: the closer we get to an understanding of the dynamism of the integral Earth, the more obvious it becomes that the four and a half billion years of terrestrial evolution resembles one vast embryogenesis. Something is developing, hatching, unfolding, and we are the self-reflexive mind and heart of the whole numinous process.

YOUTH: But what have we learned that showed that the Earth is a self?

THOMAS: We've learned, for instance, that the Earth has maintained a steady temperature over some three billion years. I say *maintained* the temperature. I mean by this the same self-organizing activity that enables the flame to maintain its process under varying conditions. The Earth has been heated by the Sun, of course, but the Sun has not remained the same temperature. The Sun's temperature has soared during the Earth's existence, rising at least twenty-five per cent of it original temperature. The Earth has responded to changes in conditions, adjusted just as the atom or tree or flame adjusts to new situations. The Earth organizes its materials so that it can maintain the narrow band of conditions enabling its life to unfold and continue.

YOUTH: How do we know the temperature on Earth has always been in that narrow band?

THOMAS: The cybernetic system of energy exchange on the Earth is extremely sensitive to any change in temperature. Even a two-degree drop in the average temperature of the Earth would lead to conditions that would freeze the Earth into solid ice.

YOUTH: But why couldn't Mars or Jupiter do the same thing?

THOMAS: They tried! Mar's evolution continued for billions of years. In the beginning, its processes in many ways resembled the Earth's, but it was unable to continue with its evolution, and has all but completely stopped. The situation is again similar to the flame: a flame will leap forth any place in the universe where a particular band of conditions is available. Put wax, oxygen, wick together under certain pressures and temperatures, and the flame will leap into action. So too with planets. A narrow band was maintained by Earth because Earth began in the right place with the right materials. Mars was close, but unable to locate itself within these conditions.

YOUTH: This is amazing. I see what you mean about the nature of this revolution.

THOMAS: For so long we've considered the Earth as just a big dead ball of dirt. It shocks us nearly out of our minds when we discover we're involved within something that *moves*. Copernicus said the Earth moved: he meant that it moved around the Sun. When we say the Earth moves, we mean the whole process is alive. The Earth moves. In a sentence that is the heart of our cosmic revolution.

YOUTH: But how does the Earth do this? How does the Earth organize itself? Where is the Earth acting to do this?

THOMAS: In your allurements, your hopes, especially in your deepest dreams for the future.

YOUTH: But how. . . .

THOMAS: The whole process is present in the individual creature. The dynamics that fashioned the fireball and the galaxies also fashion your ideas and visions. I don't mean this in a crude sense; the whole system of life and being presents itself in each particular event as well. In your specific personal dreams and desires, the whole process is present in your own personal self. The macrocosm is not disconnected from the microcosm.

YOUTH: But how? I don't see how this happens.

THOMAS: It would be impossible for you to understand easily this larger view of reality's dynamics. We have been instructed for centuries in an atomistic point of view that does not allow for the understanding of the presence of the whole in the individual. But we can consider an actual example from Earth processes that will point our attention in the right direction.

The oxygen content of our atmosphere is near twenty-one percent; this has been maintained for over a billion years. How did it get this way? The metabolic processes of the first microorganisms on the planet, the prokaryotes, added oxygen to the atmosphere and slowly increased the percentage. If these creatures could have continued unabated, the oxygen content of the atmosphere would obviously be much higher. But there came a point when the concentration of oxygen was too much for them, and they ceased being the dominant creatures of Earth. They dove to the bottoms of ponds, or hid inside other creatures. On the other hand, if these prokaryotes had been unable to continue as long as they did, the oxygen content would be much lower than it is today.

The interesting point is that the present concentration of oxygen on Earth depends on the genetic capabilities and limitations of the prokaryotes. No one came and told them to stop making oxygen when such and such a concentration was reached. They just carried on with all their delight in living until the conditions became too noxious for their own genetically anchored limitations. They organized themselves, all these little selves, within the first eons of the Earth, and continued to do just that, without any awareness of the consequences of their activities for the Earth.

But in our century we have learned something about this level of oxygen in our atmosphere. If the concentration of oxygen were increased by only several percentage points, the conditions would become such that a single lightning strike could turn an entire forest, an entire continent, into flames. On the other hand, if the concentration of oxygen were significantly lower than its present level, we would not have the large supply of chemical potential energy necessary for advanced forms of animal life. The Earth created an atmosphere that provided as much chemical potential as possible for the creation of the animal kingdom, while avoiding a situation of total terrestrial catastrophe in the spontaneous outburst of raging fires.

This is astonishing, truly; but we need in particular to think of those little microorganisms that produced the oxygen. How did they know to stop? They knew nothing of the macrostructure of the biosphere. They knew only their own allurements in the midst of their own unseen shaping powers. The whole Earth system was present in the microorganism. The macrostructure was present in the intrinsic genetic limits of the microstructure. Isn't that astonishing?

YOUTH: But how did this, how did the Earth—

THOMAS: We don't know. We are so unaccustomed to thinking in terms of whole systems that, for now, we can only guess. The point I want to emphasize is the fascinating interconnection of the whole Earth process. The prokaryotes are not separate from the atmosphere, the complex multicellular life forms, or the Earth as a self-organizing entity. I think we should take the prokaryote as the mascot of the emerging era of the Earth. What better organism to symbolize the vast intricate mystery of Earth's embryogenesis? What creature can better remind us that our own desires are rooted in Earth's desires?

YOUTH: I have to become like a prokaryote?

THOMAS: Don't say it with such a grimace! Let's just *hope* we can emulate some of the achievements of the prokaryotes!

YOUTH: In what ways?

THOMAS: To begin with, it would be wonderful if we could contribute something as essential to the Earth's life as oxygen. All the animals depend on the prokaryotes' creativity. Do you think *Homo sapiens* could match that one, or even come close to the value of our little microscopic cousins?

Secondly, we must act on our innate desires with the confidence that these are not disconnected from the Earth process as a whole. We are just now discovering a deep disgust with the industrial excesses of our consumer society. This disgust is genetically anchored, just as the cancers and other industrial diseases are genetically anchored. Our disgust and our diseases are Earth's way of making clear for us what activities are required.

Thirdly, and most importantly, we must embrace and cherish our dreams for the Earth. We are creating with our imaginations a period of rebuilding, where the intercommunion of all species will guide our life activities. We must come to understand that these dreams of ours do not originate in our brains alone. We are the space where the Earth dreams. We are the imagination of the Earth, that precious realm where visions and organizing hopes can be spoken with a discriminating awareness not otherwise present in the Earth system. We are the mind and heart of Earth only in so far as we enable Earth to organize its activities through self-reflexive awareness. That is our larger destiny: to allow the Earth to organize itself in a new way, in a manner impossible through all the billions of years preceding humanity. Who knows what rich possibilities await a planet—and its heart and mind—that have achieved this vastly more rich and complex mode of life?

WIND

THOMAS: Our last cosmic dynamic is revealed by the wind. Wind is created as heat moves from place to place. The entire universe expands in just this way: if we look into the night sky, we see that the galaxies are all moving away from us. The further away the galaxy, the faster it rushes away. This is the result of the initial explosion of the primeval fireball, when all matter was in a terrifically hot and dense form; it has all been rushing away from itself for twenty billion years.

Wind reveals the cosmic dynamic of expansion out from an area of high concentration. This dynamic creates the winds on our planet, and, in the macrostructure of the cosmos, the expansion of the universe.

YOUTH: Does this. . .dynamic have a name?

THOMAS: It's usually called the second law of thermodynamics. For instance, if you heat the center of a sheet of metal with a blowtorch, heat rushes away in all directions from the center. Heat will not stay bottled up in one spot. A similar situation in the realm of elementary particles is referred to as Pauli's Exclusion Principle: some elementary particles do not pile up on each other, but push off into separate states of being. In the biological realm, ethologists call a similar situation "dispersal behavior," where juveniles are sent out in a programmed dispersal from the occupied territory of their ancestors.

All these different words are the legacy of our splitting up of the world to study it from many points of view, a powerful analytic method to be sure, but one that makes it difficult to form a single coherent picture of the whole. We similarly fragment allurement: in the physical realm we call it "gravitation," in the biological "instinct," and in the human as "interest."

YOUTH: Is there a word that fits the human? One that points to the cosmic dynamic of wind?

THOMAS: Yes: exuberance. When you fall in love, don't you experience an irrepressible need to give expression to this joy in some way? Your outrush of poetic utterances is the human analogue to the outrush of the galaxies from a region of extreme concentration.

We can refer to the cosmic dynamic as a single activity called *celebration*. I mean in particular celebration as *announcing,* as we celebrate a new discovery in some field of science. Celebration captures the basic dynamic of expansion from a center with news of that center. The basic movement is from a region richer in being to one poorer. All the elements were at one time concentrated in the core of a star, and were then sent flooding off in all directions toward the surrounding element-poor regions of space. Young lions are born and nurtured in a particular spot on the Serengeti plains, then let loose to roam into the less populated regions. All the insights of Buddhists were concentrated on one spot of the Indian subcontinent, then released as men and women trekked into China and radiated out to Tibet and Southeast Asia, announcing the news. Superconcentration of being naturally unfolds and expands.

YOUTH: A supernova is dense in elements, I can see that. And wind requires a high concentration of heat. But what does the human have in such great concentration?

THOMAS: Being. Or, more simply, the universe. After a person has soaked in the presence of something, there is simply more there.

YOUTH: More of the universe?

THOMAS: Yes. The universe is more intensely present, desiring to explode out in celebration of itself. When you have been in the forest for a while, deepening your sensitivity to the forest's presence, the richness concentrated in you can not be contained. You radiate *forest* wherever you go, whether or not you say a single word. It would be hopeless to try and stop this natural outpouring of being, as useless as trying to stop the wind, or trying to prevent the expansion of the galaxies.

Being folds itself into concentrated fullness, then erupts in an explosion of joy. The artist sends forth her works; the parent lavishes care upon his children.

YOUTH: I love this! Is it all new?

THOMAS: Not really. The understanding of Being within the cosmic creation story, yes. But Being's innate urgency to unfold has been appreciated in different forms. Classical theologians spoke of Supreme Being's ontological desire to pour forth goodness, to share and ignite being spontaneously. They explained the human desire to share life and being as participating in Supreme Being, in Divine Reality.

YOUTH: So this desire to share and to serve as a source of goodness is real, or innate—basic. It's just the way things are, even physically.

THOMAS: Anchored in the universe, a dynamic of the cosmos.

YOUTH: We don't have to learn it? This desire to pour out goodness is just the way things are?

THOMAS: Yes, and even to this degree: perhaps the most spectacular illustration of the cosmic dynamic of celebration is the eruption of being out of sheer emptiness. Remember how elementary particles spontaneously erupt out of no-thing-ness, the ultimate realm of generation? Emptiness is permeated with the urgency to leap forth. The difficulty is with language: when we say emptiness, we fail to evoke any sense of awe for the truth of the matter.

We can use another word: the ground of being is *generosity*. The ultimate source of all that is, the support and well of being, is Ultimate Generosity. All being comes forth and shines, glimmers and glistens, because the root reality of the universe is generosity of being. That's *why* the ground of being is empty: every *thing* has been given over to the universe; all existence has been poured forth; all being has gushed forth because Ultimate Generosity retains no thing.

YOUTH: Wait a minute—I just have to stop and think about this. I have so many questions! This cosmic dynamic of celebration and generosity—we're supposed to develop this, right?

THOMAS: It was out of the dynamic of cosmic celebration that we were created in the first place. We are to *become* celebration and generosity, burst into self-awareness. What is the human? The human is a space, an opening, where the universe celebrates its existence.

YOUTH: But how are we to develop this?

THOMAS: In a sense, this power is the culmination of the others. To remember the beauty of the universe, to enhance your sensitivity to the magnificence of the Earth, to pursue the central allurement of your life; all lead to a superconcentration of being, with its inevitable desire to celebrate itself. Generosity and celebration reveal the presence of the other dynamics because they each reveal the presence of the universe in superabundant form.

YOUTH: What should I celebrate?

THOMAS: To ask that question is to weaken your powers. You never have to ask any one else what or why to celebrate; the dynamic of celebration celebrates, that's all. Self-expression is the primary sacrament of the universe. Whatever you deeply feel demands to be given form and released. Profound joy insists upon song and dance. Don't ask anyone what to celebrate: don't even ask yourself! Let celebration be. Let generosity of being happen. Nothing more is required.

Take supernovas as your models. When they had filled themselves with riches, they exploded in a vast cosmic celebration of their work. What would you have done? Would you have had the courage to flood the universe with your riches? Or would you have talked yourself out of it by pleading that you were too shy? Or hoarded your riches by insisting that they were yours and that others did not deserve them because they did not work for them? Remember the supernova's extravagant generosity and celebration of being. It reminds us of our destiny as celebration become self-aware. We are Generosity-of-Being evolved into human form.

You are the elementary particles of the fireball, elements of the supernovas, and the generosity of the ground of all being. That is your fundamental nature. Our deepest desire is to share our riches, and this desire is rooted in the dynamics of the cosmos. What began as the outward expansion of the universe in the fireball ripens into your desire to flood all things with goodness. Whenever you are filled with a desire to fling your gifts into the world, you have become this cosmic dynamic of celebration, feeling its urgency to pour forth just as the stars felt the same urgency to pour themselves out. We *know* we feel this, whereas the star simply feels it and responds.

YOUTH: But how do I know I. . . How do I know what I have to celebrate is worthwhile?

THOMAS: Every song has tremendous value! Learn to sing, learn to see your life and work as a song by the universe. Dance! See your most ordinary activities as the dance of the galaxies and all living beings. If we attempt to constrain the self-emergent expressions of joy, we bottle up the exuberance of the universe. Imagine trying to hold back a supernova! It's the same with human celebration, generosity, and creativity: try to bottle them up and you only get neurosis and destruction.

Think of the unborn of today and tomorrow, all the future generations and all the possible species. They, too, are waiting for the exuberant generosity of being. They are dependent upon it, just as you were dependent upon the generosity of the supernova five billion years ago. Fall in love, sink into intimacy with all things, explore the relationships throughout the Earth's realm, pursue your dreams, and flood all creatures with goodness.

YOUTH: I don't know whether to be excited or angry. There's so much, I'm so full of questions and plans, and I know it's going to leak away. I know I'll forget so much of this. Can you help me remember somehow?

THOMAS: We are talking about powers, and we've discussed six of them altogether: allurement, sensitivity, memory, adventurous play, unseen shaping, and celebration. That's not too much to remember, is it?

YOUTH: No, that's easy enough.

THOMAS: We've pointed out ways in which they are presented to us. That is, we looked at the night sky and reflected on allurement. We examined the seas and talked of absorption, assimilation, and sensitivity in general.

YOUTH: Yes, yes. Keep going.

THOMAS: We saw the dynamic of memory in the way the land remembers. We looked at the life forms and found there the presence of adventurous play, in exploration, free activity, and imagination. Remember the human as the baby of the universe?

YOUTH: Yes, OK.

THOMAS: Then we considered the flame, probing the meaning of the self, seeing in each of these the presence of unseen shaping. Finally, we considered wind and saw there the expansion of being, the dynamic of celebration. So: the night sky, sea, land, life forms, fire, wind. That's easy enough to remember.

YOUTH: But now I have two lists to memorize.

THOMAS: Forget them. But you'll have to remember them very carefully in order to do that. I want you to glance at the night sky and see the cosmic dynamic of allurement *intuitively.* The night sky continuously utters a single word, and that is allurement. This will have to be learned, *then* forgotten, then known. Because you were raised in the modern anthropocentric period, you have hardly ever looked at the night sky, let alone understood that the night sky speaks to you of the central dynamic of the cosmos.

In the same way, you can come to establish yourself in a relationship with the mountains so that to glance at them is to be reminded of the cosmic dynamic of memory. The mountains and the rocks shout ceaselessly: REMEMBER! Whenever water rushes over your body, it brings into your presence the reality of cosmic sensitivity and our destiny as the mind and heart of the universe. When the wind blows coolly in your face, you are feeling the activity of generosity, reminded of the great joy and destiny of celebration. And whenever you feel sunlight on your arms, you are reminded of that great cosmic flame, the unseen shaping of which permeates you and connects you to the embryogenesis of the Earth.

We need a new human in a new Earth, creating and entering new relationships with the primary realities of the universe. In the most obvious meaning, all our difficulty as a species on this planet stems from our false relationships with winds, seas, life, sunlight, and land. It's not that we're bad; we've simply been trying to live outside our true relationships with these primordial cosmic presences.

But as you move into the full universe, you will discover something stupendous. All these powers are yours! They cost nothing! They do not depend on the color of your skin, the name of your religion, or your place of birth. The further development of the Earth community depends upon our ripening as a species, but nothing is more natural for the human person to accomplish.

We sometimes fall into the delusion that power is elsewhere, that it belongs to a different group, that we are unable to find access to it. Nothing could be further from the truth. The universe oozes with power, waiting for anyone who wishes to embrace it. But because the powers of cosmic dynamics are invisible, we need to remind ourselves of their universal presence. Who reminds us? The rivers, plains, galaxies, hurricanes, lightning branches, and all our living companions.

III: THE END OF THE FIREBALL

Societal Transformation and Geological Activity

YOUTH: But we can't do this alone! We can't do this work as isolated individuals!

THOMAS: Some of the work is always done alone. But you're right. The total activity is the activity of the Earth as a whole and that includes humans. We're speaking of an enormous enterprise; one that transforms western civilization, not to mention every other culture. This monumental task is nothing less than the reinvention of the human species.

YOUTH: But who will organize this? Who will direct it?

THOMAS: It's already happening: the transformation of a society begins spontaneously, naturally. Social transformation existed for hundreds of millions of years before any human entered the life of this planet. Human societal transformation is but one example among many. We can get a clearer understanding of such change if we take the Earth's story as our fundamental context, asking our questions of it.

An ecosystem is a society. It has its own laws and citizens, its own customary interactions, its preferred species and fringe species. The whole system of life resembles a self, as we have defined self earlier. It organizes all sorts of materials, creatures, and energy into a coherent, self-sustaining process.

Consider an ecosystem in the northeastern part of North America three hundred and fifty million years ago. In particular, the tectonic plate upon which Europe's land mass rests was pressing against our own, and the pressure of this collision buckled the land, pushing up the mountains we now call the Appalachians. Why was the European plate here, and not separated from us by the Atlantic Ocean? Because there was no Atlantic Ocean then. Nor would there be any Atlantic for another hundred and fifty million years.

The Earth evolves, as do life and the stars, and the evolution of the Earth includes the movement of the continents over its surface. They collide, throw up mountain ranges, stick together for awhile, then drift apart in new directions, creating new oceans.

YOUTH: The continents move?

THOMAS: They float on the Earth's mantle, the slowest moving substance imaginable. The Earth is just the right size to keep its inner rocks in a nearly molten state. The continents actually float, sliding this way and that with cosmic patience.

The formation of mountains is so slow that the ecosystem can adjust to the changes. As climatic conditions change, the gene pools of the various regions and species change as well. Hardy cold-water bacteria, for instance, slowly become more populous in the lakes. Their own genetic types come to dominate the gene pool, where formerly this bacteria might have been a small fraction of the lake's bacterial population. Where previously they had to struggle simply to remain alive, the changing conditions—worked through natural selection—

now favor them above all. Such transformations are anchored into the genetic structures and interactions of the whole ecosystem. Thus the North American and European collision gave rise to not only a mountain range, but societal transformations as well, as the ecosystem adapts to new conditions.

The present society of the United States reflects some of these dynamics: our society is a creation of a collision between European and Native American worlds. In many ways, Europe seems the victor; but throughout the last two hundred years, the victors have been inevitably haunted by the presence of the Native American spirit. The European suspected that the native creation-centered spiritualities were necessary for true health. But only the most gifted individuals articulated this awareness. A collision that could have been worked out in psychic and spiritual space was projected out. So much of our destruction of the continent intertwines with our debilitation of women, Native Americans, and Black Americans.

Dawning in our awareness in our own decades is the recognition that the complementary interaction between these two traditions is our most significant source of social creativity and political power. We enter a period of enormous promise. The scientific-technological, Christian, masculine, individualistic, Northern European spirit joins with the ecological, animistic, feminine, communal native spiritualities in the creation of a new form of society whose significance towers over that of all other political or social events. The psychic interaction that has been proceeding in a lopsided, destructive, and unconscious fashion for five centuries has entered a new phase with the rise of the ecological, women's, Black, and Indian movements. These transformations of consciousness reestablish the balance of our psychic centers of gravity, making possible the deepest reconstruction of contemporary society. Out of the creative energy welling up from the diversity of traditions, we fashion a form of society that takes us out of a global reign of terror and into a renewed health, into a new quality of prosperity, and into a more basic delight in being human in the midst of all the life communities.

We reinvent human society by transforming our codings. This resembles the way the ecosystem rewrites its own genetic codes to transform its society. Our codes are transgenetic. Our law codes, for instance: our legal system carries on processes formerly encoded genetically in the prehuman world. Our law codes will continue to reflect this work of societal transformation, as will our education codes and processes. Traditional customs involving foods, work, and play will all bear the effects of the changes.

Our fundamental values and programs will be altered, and individuals and character types considered marginal during the past two centuries will find themselves selected by society for political power. They will be chosen where before they were passed over because they will increasingly represent the central convictions of American citizens.

YOUTH: You speak with such confidence. Are you utterly convinced this will happen?

THOMAS: Suppose you were there when the European plate began pressing against the North

American plate. These two stupendous realities begin to collide with enough momentum to continue buckling the land for another hundred million years. How difficult would it have been then to have had confidence in the appearance of mountains, and in the transformations of all the societies involved?

YOUTH: But that involves something you can *see:* the Earth's movement and all—

THOMAS: The heat in the core of the Earth and the pressure of gravitational interaction driving the collision, yes. They enable us to see the inevitability.

There is a related situation in the human world: an irrepressible emergence of human energy. I'm speaking of the energy evoked by the story of the cosmos, by the story of the galaxies and stars, the story of life and Earth. If the collision of tectonic plates gives rise to earthquakes, the emergence of the cosmic story gives rise to humanquakes. Think of it! For the first time in human history, we have in common an origin story of the universe that already captivates minds on every continent of our planet. No matter what racial, religious, cultural, or national background, humans now have a unifying language out of which we can begin to organize ourselves, for the first time, on the level of species.

All societies throughout human history have rooted themselves in fundamental stories of the cosmos. Out of their primal stories humans define what is real and what is valuable, what is beautiful, what is worthwhile, what to be avoided, what to be pursued. Modern society is no different. We too use our basic cosmology to assign power positions, making all crucial life decisions on the basis of these fundamental world views.

We are now restructuring our fundamental vision of the world. We are creating a new meaning for what we consider real, valuable, to be avoided, or pursued. The new cosmic story emerging into human awareness overwhelms all previous conceptions of the universe for the simple reason that it draws them all into its comprehensive fullness. And most amazing of all is the way in which this story, though it comes from the empirical scientific tradition, corroborates in profound and surprising ways the ecological vision of the Earth celebrated in every traditional native spirituality of every continent. Who can learn what this means and remain calm?

THE ART OF FORGING COSMIC FIRE

YOUTH: Wait a minute! I'm about to fall apart! Just what are we doing here? We're just talking, right?

THOMAS: We're just talking.

YOUTH: Well, how does it fit into everything else that's happening?

THOMAS: We're just sitting here talking, the Sun is over China, the red oak is—

YOUTH: Right, right, but what does it mean to talk? What does it really mean?

THOMAS: To understand human language, we need to place ourselves within the context of Earth as a self-organizing reality. The Earth taught itself how to create the photosynthetic processes, how to bloom forth with the power of the angiosperms, how to create topsoil; Earth did not learn these things from Mars or the Andromeda Galaxy. Earth education is self-education.

Humans are engaged in the same dynamic of self-educating reality. So, here we sit, talking, a further development of the ancient Earth activity of education. Our situation involves something new—self-reflexion—manifested especially through language, but language itself is just a part of a larger teaching process. We sit, talking, engaged in the education process of the Earth. Is that clear?

YOUTH: OK.

THOMAS: At the present moment the Earth struggles to teach itself how to become more fully self-reflexive, now carrying on activities through self-reflexive awareness that previously functioned without this quality.

YOUTH: But how does Earth educate itself?

THOMAS: Even as we speak! All of this is the Earth educating itself. Think of the language that has come alive in just this one afternoon: do you think *we* are solely responsible for that? Good heavens, no! Think of the sacrifices required of billions of creatures to make such language possible. Take a single sentence: "The fireball exploded twenty billion years ago at the beginning of time." That sentence requires nothing less than the full twenty billion years of cosmic development. It is not "my" sentence; nor does it "belong" to the theoretical scientists who first predicted the existence of the fireball, nor the experimental scientists who first detected its heat; it is a sentence of the whole Earth. Nothing less than that is required for its speaking forth. The sentence could not exist without the oceans, the rivers, the air, the life forms, and all the thousands of years of human cultural activities. Every sentence is spoken by the whole Earth. All language is spoken by the Earth as part of a biospiritual embryogenesis. You have been sitting here listening while the Earth did the talking. Language belongs to Earth as simply as the Cascade mountains belong to Earth. As Earth struggles to enter self-reflexion, you struggle to become the heart and mind of the Earth; and so Earth—including you and me—proceeds with its self-education.

YOUTH: Before this awareness slips away, tell me about words. What happens with words?

THOMAS: As you listen to this language, which is Earth's language, you become shaped by words. Your attention forms within words, your desires are shaped by words, your visions of the future are ignited by words. In all of this, the universe shapes you, shapes itself through you so that it might become more intensely present to itself through the unfurling of self-reflexive awareness.

Our primary teacher is the universe. The universe evokes our being, supplies us with creative energy, insists on a reverent attitude toward everything, and liberates us from our puny self-definition. The universe gives us fire and teaches us its use.

YOUTH: When you say "gives us fire," you mean in the sense that . . .

THOMAS: I mean in the most immediate and simple sense. The universe gives fire—real fire, the fire of the heavens.

YOUTH: How?

THOMAS: Let's start with your present moment of experience: this involves sensations, thoughts, feelings, expectations, and hopes, the whole subjectivity that defines your *now*. We think of this as the psychic manifestation of body's neurophysiological processes. Electricity flows through your nervous system in physical correlation to your subjective experiences. Do you follow that?

YOUTH: Sure.

THOMAS: So from a physical point of view, the movement of ions in your brain corresponds to your subjective experience. Different ion flows would give you qualitatively different experiences; or, equally true, a qualitatively different mood would manifest as a different movement of ions in your nervous system. The question I want to ask is simply this. What enables the ions to move? Or what enables you to think? On what power do you rely for your thinking, feeling, and wondering?

Ions don't move by their own power: they have to be pushed and tugged about. A close examination shows that an energy-soaked molecule in the brain is responsible for the ion movement. Closer examination shows that this molecule is able to push ions around because of energy it got, ultimately, from the food that you eat. The food got the energy from the Sun; food traps a photon in the net of its molecular webbing, and this photonic energy pushes and pulls the ions in your brain, making possible your present moment of amazing human subjectivity. Right now, this moment, ions are flowing this way and that because of the manner in which you have organized energy from the Sun.

But we're not done yet. Where did the photon come from? We know that in the core of our Sun, atomic fusion creates helium atoms out of hydrogen atoms, in the process releasing photons of sunlight. So, if photons come from hydrogen atoms, where did the hydrogen get the photons? This leads us to the edge of the primeval fireball, to the moment of creation itself.

The primeval fireball was a vast gushing forth of light, first so powerful that it carried elementary particles about as if they were bits of bark on a tidal wave. But as the fireball continued to expand, the light calmed down until, hundreds of thousands of years later, the energy level decreased to a point where it could be captured by electrons and protons in the community of the hydrogen atom. Hydrogen atoms rage with energy from the fireball, symphonic storms of energy held together in communities extremely reluctant to give this energy up. But in the cores of stars, hydrogen atoms are forced to release their energy in the form of photons, and this photonic shower from the beginning of time powers *your* thinking.

YOUTH: Really. . .!

THOMAS: Fires from the beginning of time empower you *right now—this instant*. What you are thinking and feeling this very moment is possible only through the cosmic fire. Your entire nervous system is rich in this fire.

YOUTH: I'm speechless.

THOMAS: You're churning with new psychic energy, aren't you? Who has evoked this energy but the universe, our primary teacher? The universe lured us into four centuries of careful empirical investigation, and now, far from finding a pile of sterile data, we are stunned to be flooded with psychic energy.

The universe bestows on us fire from the beginning of time, simultaneously evoking our profound reverence for this fire. The universe demands our response: Do we awake, dedicating ourselves to a vision of beauty worthy of our fire's origins? Do we shape this fire as it has shaped us, aware of the awesome work that has gone into providing it?

As we lie in bed each morning, we awake to the

fire that created all the stars. Our principal moral act is to cherish this fire, the source of our transformation, our selves, our society, our species, and our planet.

In each moment, we face this cosmic responsibility: to shape and discharge this fire in a manner worthy of its numinous origins. We cherish it by developing conscience in our use of it: Are we tending this fire; revering it? Are we creating something beautiful for our planetary home? This is the central fire of your self, the central fire of the entire cosmos: it must not be wasted on trivialities or revenge, resentment or despair. We have the power to *forge* cosmic fire. What can compare with such a destiny?

When I say that the universe is the principal moral authority, I mean by this the manner in which we are taught the value of the earth. The elements were bestowed on us by the stars, the complex compounds given to us by the young Earth, the informed sequences of the genes by the microorganisms, our limbs and organs by advanced life forms, and the linguistic symbols carrying our thoughts and feelings by the human venture. We could not see without the work of those who shaped the eye; could not hear without the work of those who shaped the ear. The universe created these gifts, lavishing them upon us; our first and deepest response is infinite gratitude.

That which created all of this now desires *our* creativity, commitment, and labor, *our* delight in entering with full awareness the cosmic story. The mountains and oceans, stars and life forms—all recipients of the same generosity, contributors to the unknown future culminations of our work—all tremble with the same power. Given a finite number of days in which to live, a particular store of primordial fire with which to work, who could deny that all that matters is contributing to the awesome work of fashioning the universe?

And that's why I condense our contemporary cosmological scientific story of reality by saying that the universe is a green dragon. Green, because the whole universe is alive, an embryogenesis beginning with the cosmic egg of the primeval fireball and culminating in the present emergent reality. And a dragon, too, nothing less. Dragons are mystical, powerful, emerging out of mystery, disappearing in mystery, fierce, benign, known to teach humans the deepest reaches of wisdom. And dragons are filled with fire. Though there are no dragons, we are dragon fire. We are the creative, scintillating, searing, healing flame of the awesome and enchanting universe.

ACKNOWLEDGMENTS

What was essential? Air, the foods of the Earth, rainfall of course. And the awe of summer electrical storms. But does it make any sense to list these? I think so. The stupendous silence of the mountains permeates my thinking. Shouldn't it be mentioned? Without the mystery of a forest at night I should have dried up long ago. Or the brooding that winter snows teach. My principal debt, though, is to the beauty of the night sky. And my thanksgiving is for human sentience, and its exultant expression in our scientific, artistic, and religious traditions. I can never be the same after those empowering conversations with the ICCS Spiritual Voyagers, the PLU/UPS mathematical physics group, or the Riverdale Center for Religious Research. Parents, children, families, teachers and colleagues and students; Thomas Berry; Bruce and Linda Bochte; Frank Cousens; Matt Fox; Denise Santi Swimme; and the Mysterious Source of all these realities.

The excitement of the universe expressed in such memorable phrasing...

"Seldom, if ever, within the scientific tradition, has the excitement of the universe been expressed in such memorable phrasing."

—*Creation,* May/June 1985

"The new scientific story of the universe's origin in the 'Big Bang' has been told in popular terms before, but never like this. (It succeeds) in evoking a sense of wonder about the story's music, its poetry, its meaning in everyday life."

—*San Francisco Chronicle,* October 13, 1985

"This book is worth reading—more than once."

—*Zygon,* March 1986

"This eminently readable book tells the story of the creation of the cosmos as understood by modern science. (It) is a daring attempt to transform the reader from a passive, detached consumer of information into an active, awake participant in the cosmic evolutionary process."

—*Yoga Journal,* Nov/Dec 1985

"Brian Swimme has done a service for us all by writing a book designed to blend physics and spirituality in a language accessible to young people. *The Universe is a Green Dragon* is Brian's attempt to translate complex and obscure ideas into the cadences of normal conversation. Astonishingly, it works."

—*Minding the Earth,* June 1985

"Dr. Swimme's explication of basic physical forces is mystical and ecstatic; it shocks the reader into a physiological, explosive awareness of the stuff of being."

—prominent religious educator, 1985

"Brian Swimme states it so well, gives it such depth of meaning. . .it being our place and direction in the unfolding cosmos. . .our part within the whole, both knowable and unknowable. Reading this book was an absolute joy!"

—Michael Stearns, composer and musical visionary, June 1986